Biochemistry and Human Nutrition

NIPA® GENX ELECTRONIC RESOURCES & SOLUTIONS P. LTD.
New Delhi-110 034

About the Authors

Dr. Manvesh Kumar Sihag is an Assistant Professor in Dairy Chemistry Department at Mansinhbhai Institute of Dairy and Food Technology (MIDFT), Gujarat, India. MIDFT is an initiative of Dudhsagar Dairy, Mehsana, and is provisionally affiliated with Kamdhenu University Gandhinagar, Gujarat, India. Dr. Sihag is currently working as Head, Dairy Chemistry Department since 2016 for the faculty of Dairy Science in Kamdhenu University. He obtained his Doctorate in Dairy Chemistry from National Dairy Research Institute (NDRI), Karnal, India and Masters in Food Technology from Guru Jambheshwar University of Science & Technology, Hisar, India. His research interest lies in exploitation and processing of underutilized cereals & legumes and their utilization in the development of value added/ functional foods. He is the author of more than 16 peer-reviewed journal articles, one book and five book chapters published in journals of repute. For more details, please refer: https://orcid.org/0000-0003-4896-0636

Dr. Ankit Goyal is an Assistant Professor in the Department of Dairy Chemistry at Mansinhbhai Institute of Dairy and Food Technology, Gujarat, India. Dr. Goyal acquired Bachelor degree in Medical Sciences from Kurukshetra University (2008) and Maters in Food Science and Technology (2010) from Chaudhary Charan Singh Haryana Agricultural University (CCSHAU), Hisar, India. He holds Doctorate in Dairy Chemistry (2014) from National Dairy Research Institute, Karnal, India. Dr. Goyal's research interests include development, formulation and validation of functional foods, nutraceuticals and value-added products. He is the author of more than thirty research and review articles published in peer-reviewed journals. He has also supervised two post-graduate students during their Master's degree program. For more details, please refer: http://orcid.org/0000-0003-0099-8571

Ms. Karuna Meghwal is working as Assistant Professor in the Department of Dairy Chemistry at Mansinhbhai Institute of Dairy and Food Technology, Gujarat, India. She obtained B. Tech degree in Dairy Technology from College of Dairy & Food Science Technology (CDFST), Maharana Pratap University of Agriculture & Technology (MPUAT), Udaipur (Rajasthan) in the year 2009. She completed her Masters in Dairy Chemistry from National Dairy Research Institute (NDRI), Karnal, India in 2011. She has published research and review articles in several national and international journals.

Dr. Beenu Tanwar is an Assistant Professor in the Department of Dairy Technology, Mansinhbhai Institute of Dairy and Food Technology, Gujarat, India. Dr. Tanwar acquired Master's degree in Food Technology in 2009 and Doctoral degree in Food Technology and Nutrition (2013). She is a researcher and author, having published numerous peer-reviewed articles in the fields of Food Science Technology and Nutrition. She received Junior Research Fellowship Award from University Grants Commission (UGC), New Delhi, India for her Doctoral degree program. She has attended subject-specific skill-orientated courses, training sessions, seminars, conferences, and workshops; and has received several best poster paper awards in scientific events. She has supervised nine students for their Masters' and five scholars for their Doctorate degree in the subject of Food Technology and Nutrition. For more details, please refer: https://orcid.org/0000-0002-5223-3207

Dr. Nitika Sandhu is working as an Assistant Professor in Molecular Genetics Department, School of Agricultural Biotechnology, Punjab Agricultural University, Ludhiana, India. Dr. Sandhu worked as Post-Doctoral Fellow (2014-2018) at International Rice Research Institute (IRRI), Philippines. She has been awarded Monsanto Beachell Borlaug International Scholarship for conducting Ph.D. Research. She acquired her Doctorate and Master's degree from Haryana Agricultural University, Hisar in Molecular Biology and Biotechnology. She is trained on water saving technology and breeding aspect at IRRI and physiological aspects at CIRAD France under Monsanto Beachell Borlaug International Scholarship. Dr. Sandhu has also been invited for World Food Prize ceremony and Leadership course held at USA. She is an awardee of DST Junior Research Fellowship. She has qualified CSIR-UGC NET, ASRB NET and GATE Examination. She is the author of more than 16 research and review articles published in peer-reviewed international journals with cumulative impact factor of 45. For more details, please refer: https://orcid.org/0000-0001-9793-0340

Biochemistry and Human Nutrition

Manvesh Kumar Sihag
Assistant Professor
Dairy Chemistry Department

Ankit Goyal
Assistant Professor
Dairy Chemistry Department

Karuna Meghwal
Assistant Professor
Dairy Chemistry Department

Beenu Tanwar
Assistant Professor
Dairy Technology Department
Mansinhbhai Institute of Dairy Food Technology (MIDFT)
(Kamdhenu University)
Mehsana 384 002, Gujarat, India

Nitika Sandhu
Assistant Professor (Molecular Genetics)
School of Agricultural Biotechnology
Punjab Agricultural University
Ludhiana-141004, Punjab, India

NIPA® GENX ELECTRONIC RESOURCES & SOLUTIONS P. LTD.
New Delhi-110 034

NIPA® GENX ELECTRONIC
RESOURCES & SOLUTIONS P. LTD.

101,103, Vikas Surya Plaza, CU Block
L.S.C. Market, Pitam Pura, New Delhi-110 034
Ph : +91 11 27341616, 27341717, 27341718
E-mail: newindiapublishingagency@gmail.com
www: www.nipabooks.com

For customer assistance, please contact
Phone: + 91-11-27 34 17 17
Fax: + 91-11- 27 34 16 16
E-Mail: feedbacks@nipabooks.com

ISBN: 978-81-19002-85-6

Composed and Designed by NIPA®.

Dr. Vivek Sharma
Principal Scientist
Dairy Chemistry Division
ICAR-National Dairy Research Institute
Karnal – 132001, Haryana, India

Foreword

It is great pleasure to write Foreword for the book entitled "Biochemistry and Human Nutrition". The contents of the book give a very big canvas for the students to know the basic of nutrition, as the content is precisely written as per the outlines of the course 'Biochemistry and Human Nutrition' of under-graduate level in Dairy Science and Technology.

I have gone through the whole book and have seen that the authors have divided the book into two parts: first part addresses the basic understanding of enzymes, vitamins, nucleic acids, hormones and metabolism of macronutrients. This is a very fundamental part of the book and provides the basis for the second part. The second part majorly focuses on digestion and absorption of macronutrients for different age groups, evaluation their nutritive value, allergy and hypersensitivity issues with milk consumption, and overview of planning and nutritional policies adopted by Government of India. The authors are well qualified and acquainted with the basic concepts, reasoning and knowledge to write this textbook, as indicated by their biodata, achievements and experience in the related field. All topics have been covered in a systematic manner with the clear intention to deliver the content in a very simplified form to the readers. At the end, each chapter also includes self-assessment questions to test the readers' understanding of the subject matter.

The book would also be helpful for undergraduate and graduate students of allied disciplines such as Biochemistry, Food Science, Food Science and Technology, Food and Engineering, Home Science and Nutritional Biology.

I wish all the success to the authors for the first edition of their book and initiative towards sharing knowledge about basic biochemistry, metabolism and nutrition. It is my hope and expectation that this book will provide an effective learning experience as a text-book to the undergraduate students of subject-matter and allied fields.

I wish the readers of the book a wonderful journey in reading through.

Dr. Vivek Sharma

Preface

This textbook has been written for the students of B.Tech. (Dairy Technology) course being offered by different Dairy Science Colleges and various Agricultural and Deemed Universities across the country. Though several books are already available on this subject, they are too comprehensive to be preferred by B.Tech. students. Here, this is an attempt to cover all the topics related to the course 'Biochemistry and Human Nutrition', which is written strictly based upon the course curriculum prescribed by Indian Council of Agricultural Research (ICAR), New Delhi for B.Tech. students. This book will also be helpful for those students, who study biochemistry and/or human nutrition as one of the subjects in Food Technology, Food Science and Technology, Food and Nutrition and other allied streams of under-graduate levels.

This book is mainly divided into two sections: 1) Biochemistry; 2) Human Nutrition. First section comprises eight chapters based on metabolism of macronutrients (carbohydrates, proteins and lipids), basics of vitamins, enzymes, hormones and nucleic acids. Second section deals with the digestion & absorption of macronutrients, nutritional requirements of different age groups, analytical methods for qualitative & quantitative determination of nutrients, milk intolerance & hypersensitivity, safety aspects of food additives, toxic elements and radionuclides and various nutritional polices initiated by Government of India to combat malnutrition.

The authors welcome the comments, opinions and suggestions from the faculty, students and readers so that the content of the book can be improved in a constructive way.

Authors

Contents

1

Enzyme

Introduction

Billions of chemical reactions are happening inside human body and that too with lightning speed. However, the same reactions if carried out in laboratory, they take a very long time. The reasons behind it are the biocatalysts, also known as enzymes, which are synthesized by living cells. Chemically, they are proteins (except ribonucleic acids: RNA), colloidal in nature, thermo-labile in character and very specific in their action. For example, hydrolysis of proteins by a strong acid at 100°C takes a couple of days; but same proteins can be digested within couple of hours inside human body at 37°C. Similarly, carbonic anhydrase is used as a catalyst by living cells to transfer carbon-di-oxide from cells to blood. This process is 107 times quicker than the same process if occurs without the help of enzymes. Therefore, enzymes can be defined as the biomolecules or biocatalysts which can speed up the rate of the chemical reactions. The substance upon which an enzyme acts is called as substrate and the one which is produced as the result of enzyme-substrate reaction is known as product.

Historical background

1. In 1833, Anselme Payen found that the conversion of starch into sugar can be accelerated by a compound found in malt and named it 'Diastase'.
2. Berzellius was the first scientist who coined the term 'Catalysis' in 1836. Catalysis is a Greek word which means 'to dissolve' in English (*Kata* means 'Down' and *Lyein* means 'loosen').
3. In 1857, Louis Pasteur found that lactic acid fermentation is caused by some living organisms and named them 'Ferments'.
4. In 1878, Wily Kuhne used the word 'enzyme' for the first time to indicate the catalysis taking place in biological systems. (Enzyme is again a Greek word which means *'in yeast'* in English language).

5. In 1883, Buchner isolated enzyme system from a cell free extract of yeast. He demonstrated that the 'Diastase' discovered by Payen and 'Ferments' by Pasteur were actually chemical substances, not organisms.
6. In 1926, James Sumner was the first to isolate and crystallize the enzyme 'Urease' from Jack-bean and identified it as a protein.

Nomenclature and classification of enzymes

Initially, there was no systematic pattern or nomenclature of enzymes. Some enzymes were named after the substrate they catalyze, by putting the word 'ase' after the substrate. For example, the enzyme causing hydrolysis of lactose was named as 'Lactase'. Some trivial names of the enzymes were also in fashion but trivial names such as trypsin, thrombin, pepsin etc. gave no idea about the source, function and the reaction catalyzed by the enzymes. As the enzymes are very specific in nature; there are billions of enzymes controlling different metabolic reactions in our biological system. Therefore, the need aroused for their proper nomenclature.

A nomenclature committee called Enzyme Commission (EC) was setup by International Union of Biochemistry (IUB) in 1961, which submitted its detailed report in 1964 (with the latest updates added in 1992). According to EC classification, the enzyme has two parts –first part is the name of the substrate and second is the type of reaction catalyzed by the enzyme. As per EC classification system, the enzymes are classified into six different groups conferring to the reaction being catalyzed, as shown in Fig.1.1. Every enzyme

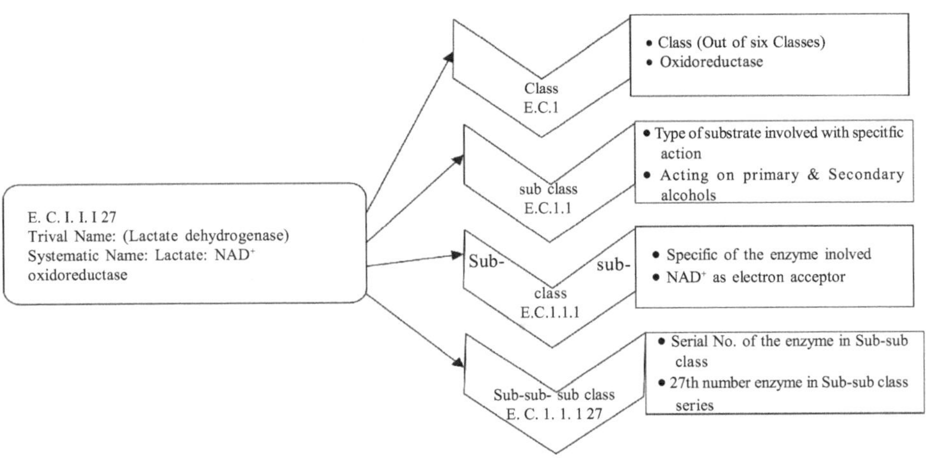

Fig. 1.1: Example of IUB classification of enzymes

had been assigned a four digit number, called as EC number. For example, if a. b.c.d are the four digits, then "a" represents the class, "b" represents the subclass, "c" represents the sub-sub-class, and "d" is the serial number of that particular enzyme in sub-sub-subclass. The "b" and "c" digits describe the reaction, while the "d" digit is used to distinguish between different enzymes of the same function. This classification system does not take into account the amino acid sequence, protein structure and chemical mechanism. The EC classification of enzymes into six different classes is shown in the Fig. 1.2.

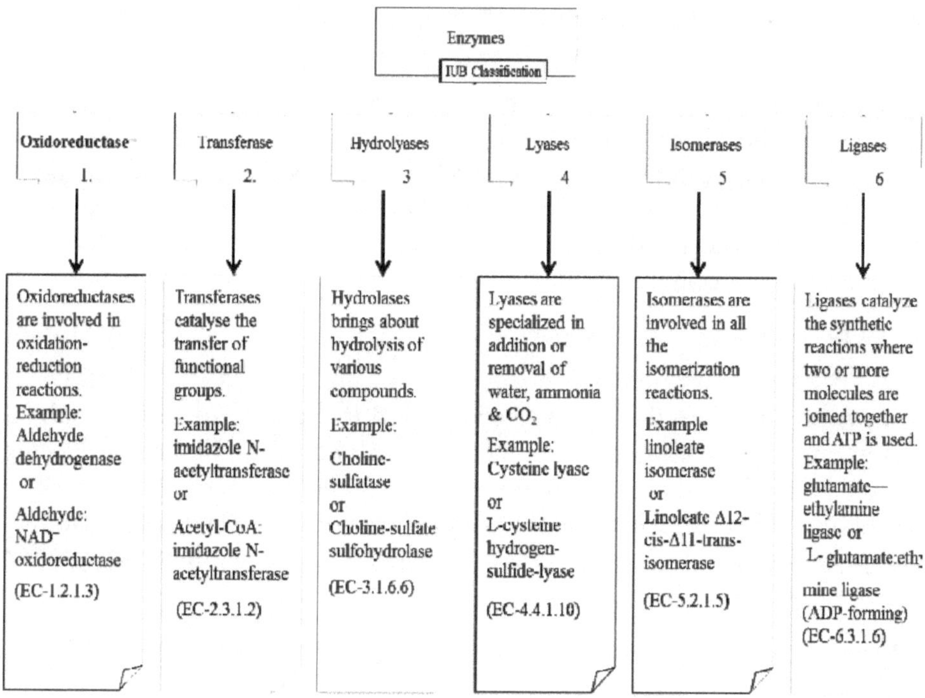

Fig. 1.2: IUB Classification of enzymes

1. Oxidoreductases: All the enzymes catalyzing oxidation-reduction reactions come under this class. The substrate which gets oxidized is regarded as hydrogen donor and which gets reduced is called hydrogen acceptor. The second digit of the four digits enzyme commission code indicates the hydrogen donor group that undergoes oxidation, means if the second digit is 1, then this indicates CHOH group undergoes oxidation, 2 indicates CHO and so on. Similarly, the third digit of the four digits code indicates type of acceptor involved i.e. if the third digit of the code is 1, it denotes NAD(P^+) is involved as acceptor, 2 indicates cytochrome, 3 indicates molecular oxygen, 4 indicates disulfide, 5 indicates quinone, 6 indicates nitrogenous group, 7 indicates iron sulfur protein and so on.

Example: Glucose oxidase (EC1.1.3.4; Systematic Name-β-D-glucose: oxygen-1-oxidoreductase).

2. Transferases: These are the enzymes which transfer a functional group (usually a methyl or glycosyl group) from a compound (donor) to another compound (acceptor). The second digit in the four digit code number indicates the group transferred i.e. 1-one carbon group, 2-aldehydic or ketonic group, 3-acyl group and so on. The third digit gives additional information of the group transferred.

Example: Formaldehyde transketolase (EC 2.2.1.3; Systematic Name: D-xylulose-5-phosphate: formaldehyde glycolaldehyde transferase).

3. Hydrolases: These are the enzymes which catalyzes the hydrolytic cleavage of carbon-carbon, carbon-nitrogen, carbon-oxygen and some other bonds. The second digit in the four digit code number indicates the nature of the bond hydrolyzed i.e.1-esterases, 2-glycosylases and so on. Third digit indicates the type of substrate i.e. in esterase whether carboxylic ester hydrolases or thiolester hydrolases or phosphoric mono ester hydrolases etc.

Example: Carboxylesterase (EC 3.1.1.1; Systematic Name-Carboxyl-ester hydrolase).

4. Lyases: These are the enzymes specialized in cleaving of carbon-carbon, carbon-hydrogen, carbon-nitrogen and some other bonds by elimination, leaving double bonds or adding/removing water, ammonia and carbon-di-oxide. The second digit in the four digit code number indicates the bond cracked i.e. carbon-carbon lyases, carbon-oxygen lyases etc. Third digit gives additional information about the group added or eliminated i.e. CO_2, H_2O etc.

Example: Carbonate dehydratase (EC 4.2.1.1; Systematic Name-Carbonate hydro-lyase).

5. Isomerases: These are the enzymes which are responsible for geometric/ structural changes within a molecule. Isomerases are involved in all the isomerization reactions. As per the type of isomerization reaction they carry out, they may be called as racemases, epimerases, isomerases (Cis-trans), mutases, tautomerases etc. The subclasses are designed as per type of isomerism and the sub-subclasses as per type of substrates.

Example: Maleate isomerase (EC 5.2.1.1; Systematic Name-Maleate cis-trans isomerase).

6. Ligases: These are the enzymes which catalyzes joining together of two molecules coupled with the hydrolysis of a diphosphate or triphosphate bond in the form of ATP. The second digit in the four digit code number represents the bond formed i.e. C-O bond, C-S bond etc. **Example:** Tyrosine- tRNA ligase [EC 6.1.1.1; Systematic Name- L-tyrosine: $tRNA^{Tyr}$ ligase (AMP-forming)].

The list of enzymes is updated regularly with some deletion of old enzymes and addition of the new enzymes. If any enzyme is deleted from the list then its number is kept vacant and not allotted to the new enzyme.

Chemical nature and properties of enzymes

The activity of an enzyme depends upon its tertiary structure and specific conformation. The functional unit of enzyme is called as **Holoenzyme** which is often further made up of a protein and a non-protein part. The non-protein part is frequently called as **cofactor** and the protein part as **Apoenzyme**.

Some enzymes do not need any additional component (cofactor) to show their full activity. However, some requires non-protein molecules (cofactors) to be bound with them for activity (Fig. 1.3). Cofactors could be either inorganic (metal ions, iron sulfur clusters) or organic compounds (flavin, haeme). The organic cofactor, if makes a tight covalent bond with the apoenzyme, is known as prosthetic group. If the organic cofactor gets released from the enzyme's active site during the reaction, then it is called as coenzyme e.g. NADH, NADPH, and ATP. The coenzyme can be removed from the enzyme through dialysis while the prosthetic group cannot be detached.

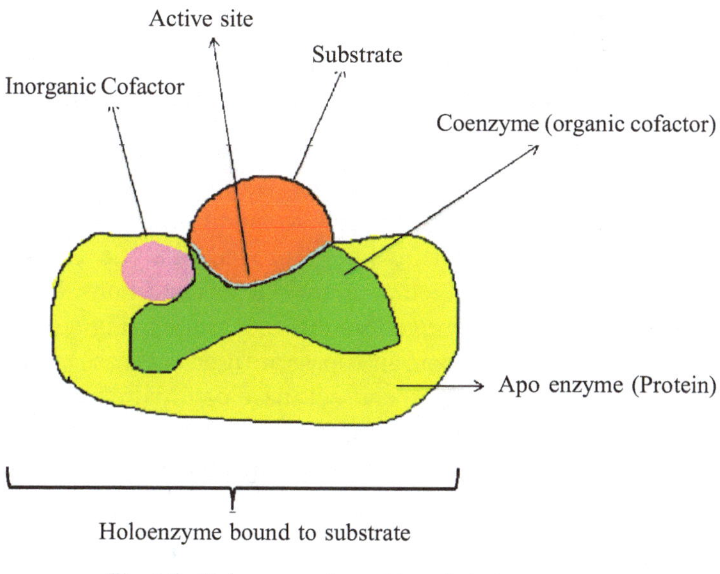

Fig. 1.3: Holoenzyme bound to substrate

Depending upon the number of polypeptide chains attached, the enzymes can be monomeric and oligomeric. For example, ribonuclease is a monomeric enzyme as it is made up of single polypeptide chain, while hexokinase and lactate dehydrogenase are oligomeric enzymes as they are made up of more than one polypeptide chain.

Multi-enzyme complex: These are the group of many enzymes having specific sites to catalyze different reactions in a definite sequence. Only the native intact multienzyme complex is functionally active and not the individual units, such as pyruvate dehydrogenase complex, fatty acid synthase complex and prostaglandin synthase complex.

Genetic engineering and modified enzymes

Enzymes are often used in molecular biotechnology for manipulation of the biological material called DNA. By genetic engineering, the DNA in an organism's genome is altered. This may be achieved by either changing certain base pairs or deleting/adding certain specific regions of the DNA. In this technique, DNA is inserted into the host genome by first isolating and copying the genetic material of interest using molecular cloning methods to generate desirable DNA sequence and then inserting it into the host genome. *E. coli* is the most preferred host for creating multiple copies of the desired DNA fragments at industrial scale due to its salient features such as well-established structure, easy to handle and faster growth rate. DNA obtained by this method is called as Recombinant DNA. Cutting and pasting DNA fragments together are the basic and fundamental techniques of Recombinant DNA technology. Restriction endonucleases, DNA ligases and a variety of DNA modifying enzymes (alkaline phosphatase, DNA polymerase, reverse transcriptase etc.) are used as a toolkit for this purpose.

By use of this technique it is possible to modify the enzymes with desirable characters and novel properties like improved catalytic activities and better performance under unusual temperature and pH conditions. Undesirable genes may be knocked out or removed by this technique using a nuclease enzyme. For example, recombinant renin and lipase are used in cheese industry, α amylase, a hybrid enzyme of glucanase and cellulase can hydrolyze barley β-glucans more efficiently during beer manufacturing. The antioxidant properties of the enzyme catalase and protease (used in detergents) are also enhanced by genetic engineering.

Site directed mutagenesis: Site directed mutagenesis or site specific mutagenesis is a useful technique for study of protein functions, identification of active sites in enzymes and designing of novel enzymes. A specific mutation at a predetermined position in a DNA molecule is produced by this technique.

This results in incorporation of a desired amino acid in place of a specified amino acid in the enzyme. *E.g. B. Subtilis* is the enzyme which is commonly present in detergents. It contains methionine amino acid which usually gets oxidized by the bleaching agent, thus inactivating the enzyme. This methionine in wild type enzyme is replaced by alanine through site directed mutagenesis making the enzyme resistant to bleach.

Factors affecting enzyme activity

(1) *Concentration of an enzyme:* Concentration of an enzyme is directly proportional to the velocity of the reaction.

(2) *Concentration of substrate:* Increase in substrate concentration gradually increases the velocity of enzyme catalyzed reaction up to a point, until the substrate is the limiting factor (Fig. 1.4). After that point reaches, further increase in substrate concentration is not going to increase the rate of the reaction any more.

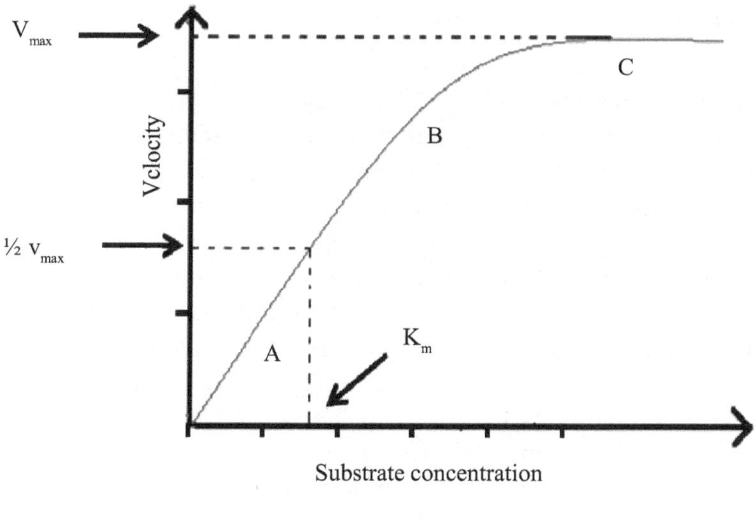

Fig. 1.4: Effect of substrate concentration on velocity

A rectangular hyperbola type of graph is obtained when velocity is plotted against the substrate concentration, having three distinct phases (A- Linear, B- Curve & C- Almost unchanged). Phase [A] is the maximum activity region as there is equilibrium between enzyme-substrate complex formation and breakdown of this complex. In phase [B], the ES complex concentration is constant but end product and substrate concentration are changing. In last phase [C], the reaction rate remains same or slightly decreases despite excess of substrate.

Enzyme kinetics and K_m value:

$$E+S \underset{K_2}{\overset{K_1}{\rightleftharpoons}} ES \xrightarrow{K_3} E+P$$

The enzyme (E) and substrate (S) combine with each other to form an unstable enzyme-substrate complex (ES) which ultimately breaks down to form the product (P).

Here K_1, K_2 and K_3 represent the velocity constants for the respective reactions and can be represented by a single constant called K_m, the Michaelis-Menten constant. This can be represented by the formula

$$K_m = \frac{K_2+K_3}{K_1}$$

K_m or **Michaelis-Menten constant** may be defined as the substrate concentration to produce half-maximum velocity in enzyme catalyzed reaction. It specifies that when the substrate concentration equals the K_m value, 50% of the enzyme molecules are bound with the substrate molecules.

(3) Effect of temperature: The kinetic energy of the molecules increases by increase in temperature, which in turn increases the random collisions between enzyme and substrate molecules per unit time, thus, ultimately increasing the rate of the reaction. However, there is also an increase in the vibrational energy of the molecules which leads to strain on the secondary bonds of the enzymes that holds them together. Further increase in temperature results in breaking of hydrogen bonds followed by ionic bonds. As a result of which the shape of the active site of an enzyme changes and the substrates are no more capable of getting fit in to the active site. Hence, the rate of the reaction decreases. If still the temperature continues to rise, then the enzyme gets denatured and the enzyme activity stops completely (Fig.1.5). Velocity of an enzyme catalyzed reaction surges with upsurge in temperature up to a maximum and then declines and a bell shaped curve is obtained. For every 10°C rise in temperature from 0°C to 40°C, the rate of reaction becomes almost double. Most of the enzymes in humans are optimally active at 37°C.

Temperature coefficient or **Q10** is defined as increase in enzyme velocity when the temperature is increased by 10°C. The optimum temperature range for most of the enzymes is 40-45°C. However, a few enzymes are active even at 100°C such as venom phosphokinases. Some plant enzymes have optimum activity at 60°C such as urease. This may be due to highly stable structure & conformation of these enzymes.

(4) Effect of pH: Whether a solution is acidic or basic is decided by its pH. The pH also gives an idea about the concentration of H^+ and OH^- ions in a solution. At pH 7, the solution becomes neutral. Below this pH, the solution contains higher concentration of H^+ ions, therefore, is acidic. Above pH 7, there are more OH^- ions in the solution and it is basic (alkaline). When an enzyme interacts with a substrate in a solution, then H^+ and OH^- ions of the solution may interact with the hydrogen and ionic bonds of the enzymes. This interaction may lead to the change in shape of the enzyme molecules, particularly the active sites.

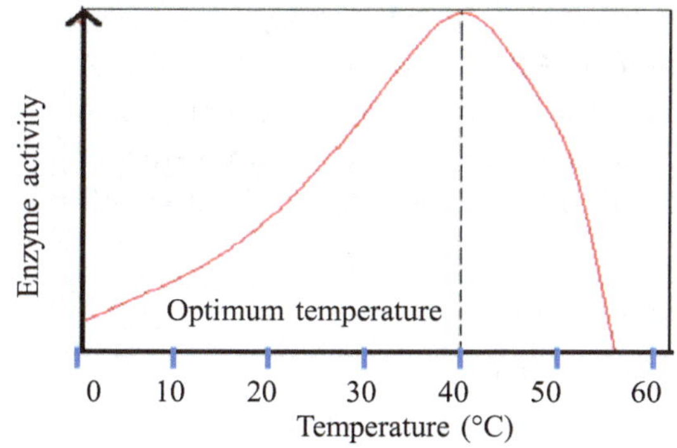

Fig. 1.5: Effect of temperature on enzyme activity

The pH at which the hydrogen and ionic bonds of the enzyme molecule either do not react with the ions or react in such a complementary way that the active site of the enzyme molecules becomes most appropriate for binding with the substrate, that pH is called as optimum pH for that enzyme. The activity of the enzyme decreases below or above this pH. Here also a bell shaped curve is obtained as shown in the Fig. 1.6.

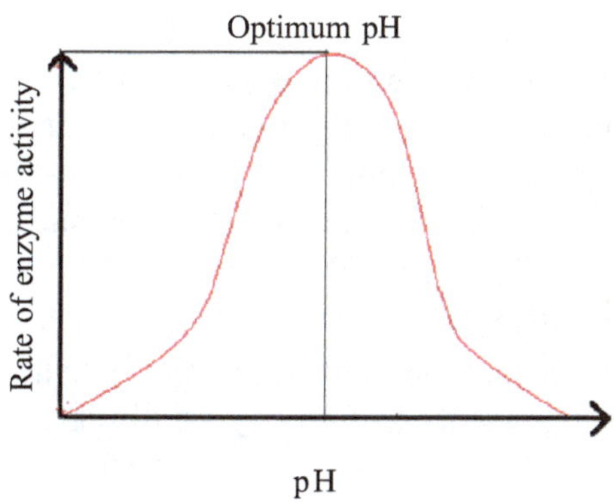

Fig. 1.6: Effect of pH on enzyme activity

Most of the enzymes of complex creatures (higher animals) show optimum activity around neutral pH (6.8). However, there are many exceptions such as pepsin (pH 1-2), acid phosphatase (pH 4-5) and alkaline phosphates (pH 10-11). Hydrogen ions may also alter the ionic charges on the amino acids, substrate and enzyme-substrate complex; hence, influence the enzymatic activity.

(5) *Effect of product concentration:* The velocity of an enzyme catalyzed reaction generally decreases as the reaction products accumulate. Sometimes, the products combine with active site of enzymes to form a loose complex resulting in inhibition of enzymatic activity.

(6) *Effect of activators and inhibitors:* Some enzymes require inorganic cofactors like metal ions (Mg^{2+}, Mn^{2+}, Zn^{2+}, Ca^{2+} and Co^{2+}) for their optimum activity. Metal ions enhance the rate of enzymatic reaction through various mechanisms like combining with the substrate, formation of E-S-metal complex, direct participation in the reaction or bringing about a conformational change in the enzyme.

There are two categories of enzymes requiring metals for their activity:

Metal activated enzymes: Metal is not tightly held by the enzyme and can be exchanged easily with other ions such as ATPase (Mg^{2+} and Ca^{2+}), Enolase (Mg^{2+}).

Metalloenzymes: Metal ions are held tightly and are not exchanged with other molecules. For examples: alcohol dehydrogenase, carbonic anhydrase.

(7) *Effect of light & radiations:* Exposure to UV light, beta-rays, gamma-rays and X-rays inactivates certain enzymes due to formation of free radicals and peroxides. For example, UV-rays inhibit salivary amylase activity.

(8) *Effect of Pressure:* Enzyme activity can also be inhibited by application of high pressure. The tertiary structure and hydrogen bonds responsible for holding secondary structure of proteins gets disrupted at pressure above 1200 MPa. The water molecules are pressed in to the interior hydrophobic region due to excess pressure and this changes the hydration of enzymes. Therefore, the enzymes can be denatured by applying pressure above 300 MPa at room temperature, resulting in distortion in their structure and activity.

Active sites of enzymes

Enzymes are big in size compared to substrate molecules. Active site (a three dimensional entity) represents a relatively small region on enzyme's surface responsible for binding with the substrates and cofactors, if any. This also contains catalytic groups which take part in formation and breakdown of bonds with substrate molecules. Although there are millions of enzymes, each possessing active sites, but some features of the active sites can be generalized as mentioned below.

Salient features of active site

1. Active site is made up of small number of amino acids which are precisely placed in three dimensional spaces such that, their side chains craft the chemistry required to catalyze a substrate. These amino acids are sterically constrained and are distant from each other in their linear sequences that make up the protein.
2. Active sites are considered as clefts or crevices or pockets conquering a small region in a big enzyme molecule.
3. Active site is flexible in structure & shape to facilitate the specific substrate binding.
4. Active site is made up of a substrate binding site and a catalytic site.
5. The cofactors on which some enzyme depends are a part of catalytic site of the active site.
6. Enzymes binds with their substrates through a system of weak, non-covalent intermolecular interactions such as hydrogen bonding, hydrophobic and electrostatic interactions.
7. Serine, aspartate, histidine, cysteine, lysine, arginine, glutamate and tyrosine are the commonly found amino acids at the active site. Out of these, serine is the most frequently found amino acid.
8. Enzymes are specific with respect to their functions and substrates because of the active sites.

Enzyme inhibition

An enzyme inhibitor is a substance or molecule which binds with an enzyme and interacts with it in such a way that prevents it from working in normal fashion. Inhibitors can be organic or inorganic in nature and specific or non-specific in their action. Non-specific inhibitors affect all the enzymes in the same way. Temperature, acid and base are some of the examples of non-specific inhibitors as they bring about chemical or conformational changes in the enzymes, which finally denature the protein portion of enzyme molecules. There are three main categories of inhibition by specific inhibitors.

(1) Reversible inhibition

(2) Irreversible inhibition

(3) Allosteric inhibition

(1) Reversible inhibition: In this type of inhibition, inhibitor binds non-covalently with an enzyme. When inhibitor is removed, enzyme inhibition can be reversed. Reversible inhibition is further of two types:

(a) Competitive inhibition

(b) Non-competitive inhibition

(a) Competitive inhibition: In this type of inhibition, the inhibitor closely resembles in molecular geometry and chemical structure with the real substrate and therefore, is regarded as substrate analogue. Inhibitor contests with substrate and binds at the active site of the enzyme but does not undergo catalysis. As long as the active site of the enzyme is blocked by the competitive inhibitor, the enzyme is unavailable for binding with substrate. Therefore, enzyme inhibition is subjected to inhibitor concentration, substrate concentration, and the comparative affinities of the inhibitor and substrate for the active site. These types of inhibitors are also often used in medical science. For example, nitric oxide binds with the receptors present in the smooth muscle cells of penis resulting in increased levels of cyclic guanosine monophosphate (cGMP) which further increases vasodilation. Phosphodiesterase type 5 (PDE-5), an inhibitor, degrades this cGMP, restricting vasodilation of penis arteries. A drug called 'Sildenafil' fits into the same active site of PDE-5 as cGMP, thus competitively inhibiting PDE-5 from working and corrects the erectile dysfunction.

Similarly, Penicillin also acts as competitive inhibitor and inhibits a bacterial enzyme which is responsible for forming cross-links in bacterial cell wall, thereby, stopping reproduction and infection.

(b) Non-competitive inhibition: In this inhibition, a site other than the active site on enzyme surface is occupied by the inhibitor, resulting in impairment in enzyme functionality. There is no structural resemblance between inhibitor and the surface of the substrate. The inhibitor also does not interfere with the enzyme substrate binding (Fig. 1.7). However, the catalysis is prevented by distorting the enzyme conformation. Non-competitive inhibitors are not affected by the substrate concentration. Heavy metal ions such as Ag^+, Pb^+ and Hg^+can non-competitively inhibit the enzymes by binding with cysteine sulphydryl groups. Cyanide is an example of non-competitive inhibitor as it combines with a cytochrome enzyme called dehydrogenase, responsible for the transfer of hydrogen atoms during cellular respiration.

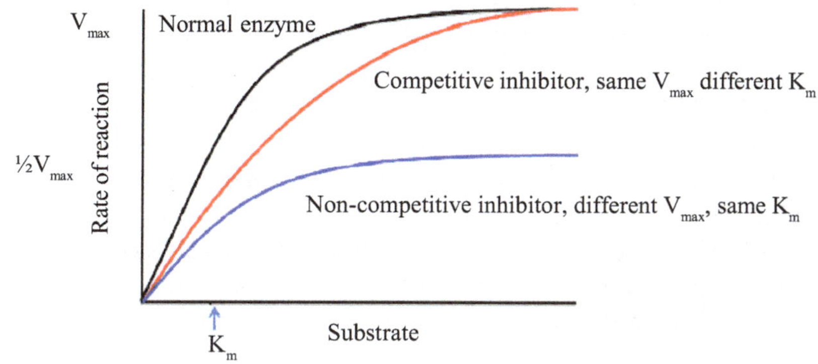

Fig. 1.7: Comparison between competitive and Non-competitive inhibition

(2) Irreversible inhibition: In this type of inhibition, inhibitor binds either with enzyme or enzyme-substrate complex to form enzyme-inhibitor complex or enzyme-substrate-inhibitor complex, which ultimately results in formation of covalently modified 'dead-end complex'. These inhibitors are usually toxic compounds. For example, di-isopropyl fluoro-phosphate (DIPF) is a nerve gas. This gas acts as irreversible inhibitor to enzyme acetyl cholinesterase. Acetylcholine esterase causes hydrolysis of acetylcholine. Acetylcholine is the neurotransmitter used at the neuromuscular junction. DIPF is capable of forming covalent bonds with the hydroxyl group of serine amino acid on active site of this enzyme inhibiting the enzyme activity irreversibly. This inhibition increases both the level and duration of action of the neurotransmitter acetylcholine

Similarly, Tosyl-L-phenylalanine chloromethyl ketone (TPCK), acts as chymotrypsin analog and binds with the histidine amino acid at the active site of the enzyme, inhibiting the enzyme activity. Triose phosphate isomerase, also an irreversible inhibitor, resembles with the substrates in structure and binds with the active sites of the enzymes covalently making them inactive.

(3) Allosteric inhibition: Besides the active sites, some enzymes possess additional sites known as allosteric sites. These types of enzymes are called as allosteric enzymes. Certain substances known as allosteric modulators bind at the allosteric site and regulate the enzyme activity. The activity is increased when a positive allosteric effector binds with the allosteric site; and activity is decreased when a negative allosteric effector binds at the allosteric site. For example, strychnine, a highly toxic alkaloid acts as allosteric inhibitor and binds to glycine receptors. This binding prevents glycine (an inhibitory neurotransmitter) from attaching with glycine receptors. This causes motor neurons to nonstop fire, and the patient suffers constant muscle convulsions and eventual death through asphyxia. Similarly, 2, 3-Bisphosphoglyceric acid (2.3 BPG) acts as negative allosteric effectors to hemoglobin and decreases its oxygen carrying capacity significantly.

Enzyme specificity

Specificity may be called as ability of an enzyme to identify and react with the exact substrate among a group of several similar chemical molecules. Enzymes are highly specific in their action when compared with the chemical catalysts and that's why there are millions of enzymes in biological system. This property of the enzymes is also a very useful diagnostic and research tool. Enzyme specificity is mainly of three types:

(1) Stereo-specificity or optical specificity

(2) Reaction specificity

(3) Substrate specificity

(1) Stereo-specificity: This is considered as highest class of specificity by any enzyme in living systems. Stereo-isomers are those compounds which have same molecular formula but different structural configuration. The enzyme acts only on one isomer and therefore, shows stereo-specificity. L-amino acid oxidase and D-amino acid oxidase acts on L and D amino acids, respectively. Similarly, β-glycosidic bonds of starch and glycogen are hydrolyzed by α-glycosidases only and β-glycosidic bonds of the cellulose by β-glycosidases only.

(2) Reaction specificity: When the same substrate undergoes different type of reactions and each reaction is catalyzed by a separate enzyme, this is called as reaction specificity. For example, an amino acid can undergo transamination, oxidative deamination, decarboxylation, racemization, etc. in presence of different specific enzymes.

(3) Substrate specificity: Substrate specificity can be divided further into three types:

(a) ***Absolute substrate specificity:*** Certain enzymes act only on one substrate. Lactase can hydrolyze only β-1-4 linkage of lactose. Glucokinase acts only on glucose to give CO_2 and glucose-6-phosphate. Similarly, urease cleaves urea to produce ammonia and CO_2.

(b) ***Relative substrate specificity:*** Some enzymes acts on structurally related substances and this type of action may be due to a specific bond or group surrounding the bond.

(i) ***Group specificity***: This is also called as moderate specificity. Pepsin cleaves peptide bonds contributed by aromatic amino acids like phenylalanine, tyrosine, and tryptophan. Similarly, trypsin cleaves the peptide bond contributed by any basic amino acid such as histidine, lysine and arginine.

(ii) ***Bond specificity***: Glycosides act on glycosidic bonds of carbohydrates. Similarly, lipases act on ester bonds on lipids.

(c) Broad specificity: Some enzymes act on closely related substances, thus, show broad specificity. For example, hexokinase can act on glucose, fructose, mannose and glucosamine also.

Mechanism of enzyme action

For catalyzing a reaction, the substrate must combine with the enzyme at its active site and form an enzyme-substrate complex (Fig. 1.8), which ultimately breaks down to yield product. Three theories have been put forward to explain this mechanism of enzyme-substrate complex formation:

(a) Lock and key theory: This theory was given by a Germen scientist Emil Fischer in 1894 and is an old model. Therefore, it does not fit to all enzymes. According to this theory, the structure or conformation of an enzyme is rigid and the substrate fits in to the active site of the enzyme just as a key fits into a proper lock or a hand into a proper glove. The side chain (R-group) of the amino acids help in mediating the interaction between enzymes active site and the substrate.

Drawback: This theory is generally applicable to only those enzymes that catalyzes only single type of substrate and does not support the flexible nature of enzymes and hence fails to explain the effect of allosteric modulators.

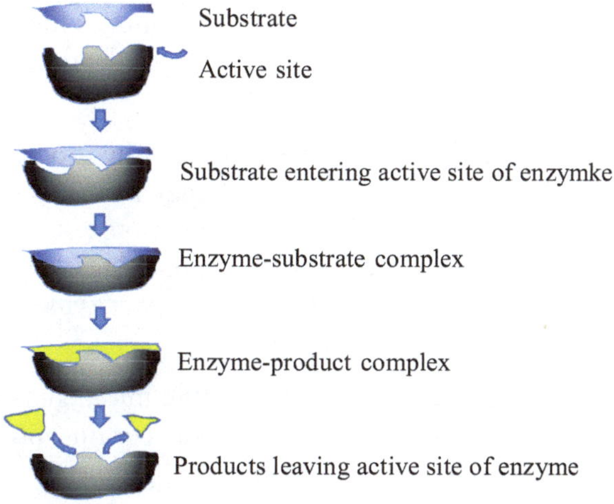

Fig. 1.8: Enzyme substrate complex formation

(b) Induced fit theory or Koshland's model: This theory was proposed by Koshland in 1958 and was comparatively more acceptable and realistic model for enzyme-substrate complex formation. According to this theory, the active site is not rigid and substrate plays a role in enzyme's final shape determination. When substrate reacts with the enzyme, it prompts a conformational modification in the enzyme, resulting in the construction of a strong substrate binding site. Further, due to induced fit mechanism, the appropriate amino acids of the enzyme are repositioned to form the active site and bring about the catalysis.

Advantage: This theory explains the action of allosteric modulators and competitive inhibitors on enzymes. It also explained why certain molecules can bind with the enzymes but do not undergo catalysis, as the enzyme may be distorted too much by those molecules.

(c) Substrate strain theory: According to this theory, when substrate binds to the active site, the enzymes induce a strain to the substrate. This strained substrate leads to formation of product.

In fact, a combination of the induced fit model with the substrate strain is considered to be operative in the enzymatic action.

Regulation of enzyme activity in living system

There are millions of cells of different types in human body and each cell contains thousands of different enzymes. As different cells need different enzymes at different time; therefore, all the enzymes are not active or functional all the time. For example, digestive enzymes are active when there is food in the stomach. Similarly, there is a variation in needs and conditions of skin cells, bone cells, blood cells, nerve cells and muscle cells for enzymes. Therefore, human body beautifully controls or regulates the amount and functionality of different enzymes in different cells at different times through one of the pathways mentioned below:

(1) Allosteric or feedback regulation: The process of inhibiting the first step by the final product in a series of enzyme catalyzed reactions of a metabolic pathway is known as feedback regulation. As the metabolic demand for the final product decreases, it starts accumulating in the surroundings of the reaction and eventually binds with the regulatory enzyme at the start of the sequence of reaction, inhibiting the enzyme.

For example, here in Fig. 1.9, A is the initial substrate; B, C, D and F are the intermediates. E & G are the end product or final product in a pathway catalyzed by six different enzymes (e_1, e_2, e_3, $e_{4,}$ g_1& g_2).

The very first step (A to B by enzyme e_1) is most effective for regulating the pathway by the final end product E. This type of control is often called negative feedback control or regulation.

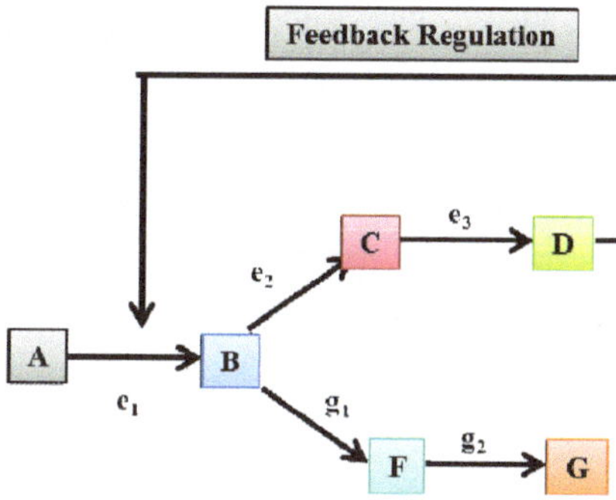

Fig. 1.9: Feedback regulation

(2) Activation of latent enzymes: Latent enzymes are those enzymes which remain inactive and a small part of their polypeptide chain needs to be removed in order to show enzymatic activity. Most of the digestive and blood clotting enzymes are latent enzymes. These enzymes are synthesized as Pro-enzymes or Zymogens, which undergo irreversible covalent activation by the breakdown of one or more peptide bonds, as and when they are needed by our metabolic system. For example, pro-enzymes mainly chymotrypsinogen, pepsinogen and trypsinogen are converted into the active chymotrypsin, pepsin and trysin, respectively, as the food passes in our digestive track. Similarly, some enzymes in our body exist in the active and inactive forms which are inter-convertible depending on the needs of the body. This inter-conversion is brought about by the reversible covalent modifications, namely either phosphorylation and de-phosphorylation or oxidation and reduction of disulfide bonds. For example, pyruvate dehydrogenase exists in to inter-convertible active and inactive forms. Pyruvate dehydrogenase remains in inactive form which is converted by dephosphorylation into active form as and when required by the body (Fig. 1.10).

(3) Compartmentation: There are certain substances in the body which are synthesized and also degraded at periodic intervals. Generally the synthetic and breakdown reactions are operative in different cellular pathways to achieve maximum economy. For example, enzymes for fatty acid synthesis are found in cytoplasm whereas enzymes for fatty acid oxidation are present in mitochondria. So, depending upon the needs of the body, fatty acids are either synthesized or oxidized.

(4) Control of enzyme synthesis: There are two types of enzymes: Constitutive enzymes or House-keeping enzymes and Inducible enzymes. The level of constitutive enzymes is not controlled and remains almost constant. The concentration of inducible enzymes decreases or increases as per the body needs and are well regulated by induction or recession mechanism.

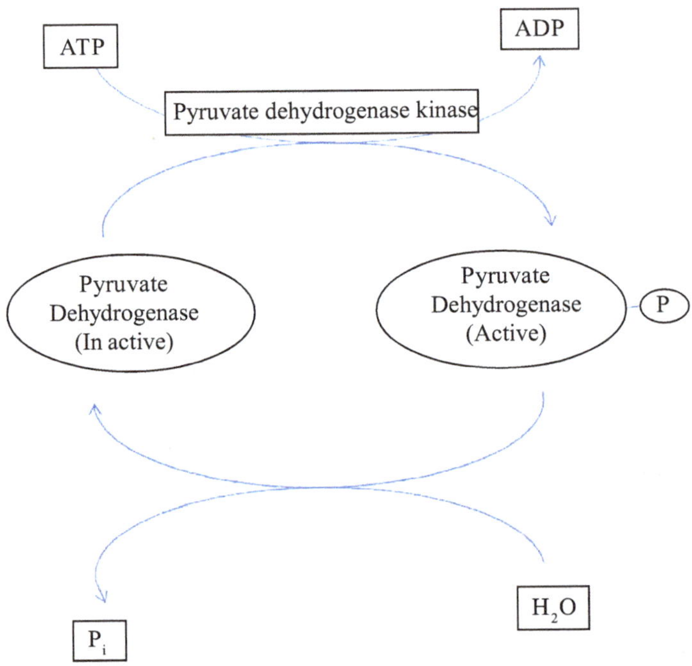

Fig. 1.10: Activation of latent enzyme Pyruvate dehydrogenase by dephosphorylation

Induction means increased synthesis of enzymes in response to some specific molecules (inducer molecules) and repression means decreased synthesis, again in response of some molecules called as repressor molecules. For example, when bacteria are grown on glucose media, then they act as an inducer for lactase enzyme. Similarly, the hormone insulin induces the synthesis of glycogen synthetase, glucokinase, phosphofructokinase and pyruvate kinase. All these enzymes are involved in utilization of glucose.

Pyruvate carboxylase is a key enzyme for the synthesis of glucose from non-carbohydrate sources like pyruvate and amino acids, but if sufficient glucose is available there, then there is no necessity for its synthesis. So, glucose causes repression of pyruvate carboxylase. The rate of synthesis of α-hydroxy, α-methyl glutaryl Co-A reductase is a key enzyme in cholesterol biosynthesis. When there is enough cholesterol in diet, then the dietary cholesterol acts as repressor for this enzyme.

(5) Enzyme degradation: Enzymes are not immortal yet there is a lot of variation in half lives of individual enzymes. Half-life for some enzymes is in days while for others in hours or even minutes. Lactate dehydrogenase (LDH_4) half-life is 5-6 days, LDH_1 8-12 hours and amylase 3-5 hours.

(6) Isozymes: Multiple forms of the same enzyme also help in regulation of enzyme activity.

Units for measuring enzyme activity

As enzymes are partially active and their purity is not much, therefore, it is difficult to determine the amount of enzyme in absolute terms to be used for a reaction. Hence, enzymes are expressed not in terms of their concentration rather in terms of their activity. Previously, non-standard enzyme activity units such as Soxhet, Anson and Kilo Novo were used, which were based on physical changes such as lowering viscosity. Some other units of enzyme activity are King Armstrong unit, Somogyi unit, Reitman frankel unit, spectrophotometric units, etc. Now-a-days, 'Katal' and 'International unit' are commonly used for the determination of enzymatic activity.

Katal: In order to maintain the uniformity in the expression of enzyme activities, a new unit Katal (abbreviated as Kat) was introduced by IUB. One Katal (kat) denotes the amount of enzyme which catalyzes the transformation of one mole of substance per second (1 kat = 60000 000 IU).

International unit (IU or U): One SI unit or IU is defined as the amount of enzyme activity that catalyzes conversion of one micromole of substrate per minute.

Ribozyme

Ribosomes are the exceptional enzymes in terms of the fact that they are actually not protein in nature. Before the detection of ribozymes, enzymes were anticipated to be the only catalytic proteins. Ribozymes were first discovered in 1980 by Thomas R. Czech, who was awarded Nobel Prize in Chemistry in 1989 for this work. A ribozyme is a RNA molecule that catalyzes a chemical reaction. It increases phosphate diester bond cleavage and peptide bond synthesis rate. Ribozyme molecule consists of a protein and a non-protein part. The non-protein part (RNA) accounts for 80% and protein part for 20% of the total ribosomes. Sidney Altmann and co-workers in1983 isolated RNA part from the ribonuclease P and it showed true enzymatic activity and also obeyed Michelis-Menten kinetics. Therefore, they concluded that ribonuclease P was functional due to RNA component present in the enzyme and not the protein part of the enzyme. Many naturally present ribozymes catalyze either the hydrolysis of one of their own phosphodiester bonds or the hydrolysis of bonds in other RNAs. They had also been found to catalyze the amino transferase activity of the ribozymes.

The reason for showing their catalytic activity may be that they can form complicated secondary and tertiary structure just as in the case of proteins. Although most ribosomes are quite rare in cells, yet there roles are essential to life. For example, the functional part of ribosomes, the molecular machine that translates RNA into proteins, is basically a ribozyme.

Immobilized enzymes

Enzyme immobilization may be defined as binding the enzyme molecules to a solid insoluble matrix which does not affect the enzyme stability or its catalytic activity. The enzyme molecules are confined with a solid support called matrix or their movement is restricted somehow and substrate is allowed to pass over the matrix for converting it in to the product.

Matrix used for immobilization could be inorganic (porous glass, silica, celite, centonite), organic natural (DEAE cellulose, CM cellulose, dextran, chitin, chitosan) or organic synthetic (polystyrene, polyvinyl acetate, acrylic polymers) in nature; and should possess following properties:

- Low cost
- Inertness
- Good physical strength
- Good stability
- Regeneration of the matrix
- Enhancement of enzyme specificity
- Reduction in product inhibition

For example, glucose oxidase and peroxidase immobilized and coated on a strip of paper are used in the clinical laboratory for detection of glucose in urine.

Benefits of immobilization

- Enzymes get protected from degradation and deactivation and it also provides enhanced stability to them.
- Easy separation of enzymes from the reaction mixture.
- Reuse of enzymes for many reaction cycles, lowering the total product cost

- Ability to control the rate of the reaction at specific point
- Enhanced stability
- Ability of enzymes to replace multiple standard chemical steps and provide commercially pure and safe product as immobilized enzymes gets easily separated from the product.

Methods of immobilization

The enzymes can be immobilized by either physical methods or chemical methods. Physical and chemical methods are further classified as shown in Fig. 1.11.

(1) Physical adsorption on an inert carrier: This is a very simple method with wide applications and capable of high enzyme loading (about one gram per gram of matrix) capacity. In this method, enzyme is simply mixed with an adsorbent, under appropriate conditions of temperature, pH and ionic strength; followed by an incubation period and finally washing off the unbound or loosely bound enzyme. This binding of enzyme with adsorbent is usually due to weak interactions like vander-waal forces of attraction or hydrogen bonds. For example, silica, bentonite, cellulose etc. are the common matrix and the enzymes immobilized by this method are catalase and invertase.

Advantages: Simple and economical method with very limited loss of activity. Enzymes can be recycled, reused and regenerated.

Disadvantages: This method offers low surface area for binding compared to other methods. Enzymes are exposed to microbial attack and product yield is low.

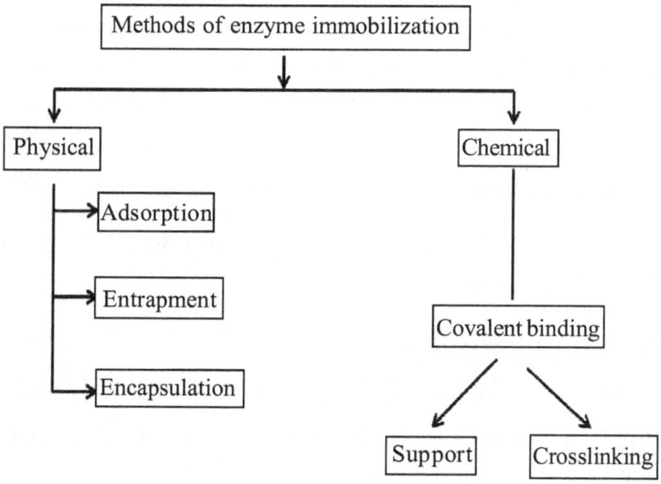

Fig. 1.11: Methods of Immobilization of enzymes

(2) Entrapment and encapsulation: In this type of immobilization, the enzymes are not directly bonded with the support as they are simply trapped inside suitable gels (poly acryl amide gel) or fibre network (cellulose) or microcapsules (polyamine) of a polymer matrix. If the enzymes are entrapped within the interspatial spaces of a cross linked polymer of hydrophilic nature, then it is called as lattice type entrapment; and if the enzymes are enclosed within semi-permeable collodion (a syrupy solution of nitrocellulose) or nylon membrane, then it is called microcapsule type entrapment.

Advantages: The enzymes are relatively stable with no chemical modification, easy to handle and reusable.

Disadvantages: There are chances of enzyme leakage from the pores of the lattice or microcapsules.

(3) Covalent binding to a reactive insoluble matrix/support: In this method, small amount of enzyme is immobilized by formation of covalent bonds with water soluble carriers by the help of some functional groups such as amino group, carboxyl group, sulfahydryl group, hydroxyl group, thiol group, etc. The main disadvantage of this method is that the conformational structure of the enzymes may get altered by covalent binding resulting in low enzyme activity or enzyme inhibition. However, as the bonding involved is of strongest type in nature therefore, the chances of enzyme leakage are not there. Enzyme loading capacity of the method is about 0.02g per gram of matrix. Lysine amino acids are very useful in covalent binding of enzymes with insoluble matrix due to their wide spread surface exposure and high reactivity especially in alkaline medium. Lysine also appears to be very rarely present at the active sites of the enzymes.

(4)Cross linking: In this type of covalent binding, enzyme molecules form intermolecular cross linkages with a solid support resulting in formation of a three-dimentional cross linked enzyme aggregates with the help of multifunctional reagents.

Advantage: Low desorption and comparatively higher enzyme stability.

Disadvantage: This is not a cost effective method and there are chances of conformational change in enzyme's structures.

Limitations of enzyme immobilization

- Carriers and the immobilization process both are usually costly.
- There may be changes in enzymes with respect to their selectivity and specificity.
- Many enzymes need cofactors for their performance so regeneration of these cofactors is also a technical issue.

- If a reaction is controlled by multi-enzyme complex, then to immobilize all enzymes to a single matrix or using different matrix at proper interval is also problematic.
- There are chances of loss of enzyme activity during immobilization as the active site of the enzyme may get involved in binding with the matrix.

Sample questions

1. Write a short note on nomenclature and classification of enzymes.
2. What was the need of IUB classification of enzymes?
3. Write a short note on chemical nature and properties of enzymes.
4. How the enzymes can be manipulated by genetic engineering techniques?
5. Briefly describe the various factors affecting enzyme activity.
6. What do you understand by active site of an enzyme? Briefly explain its salient features.
7. Discuss enzyme inhibition in detail citing examples.
8. What do you understand by enzyme specificity? Discuss various types of specificities shown by enzymes in brief.
9. Discuss the main theories prevailing about mechanism of enzyme action.
10. Explain how human body regulates the activities of various enzymes?
11. Write a short note on Ribozymes.
12. What is enzyme immobilization? Briefly discuss its techniques, their advantages and limitations.

Suggested Readings

Boyer, P.D. and Krebs, E. G. (1987). The Enzymes: Control by Phosphorylation, Part B: Specific Enzymes.

Cox, M.M. and Nelson, D. L. (2008). Lehninger Principles of Biochemistry. 5th edn. New York: Worth Publishers.

Eskin, N.M.A. and Shahidi, F. (2013). Biochemistry of Foods. 3rd edn. USA: Academic Press Elsevier.

Krajewska, B. (2004). Application of chitin-and chitosan-based materials for enzyme immobilizations: a review. *Enzyme and Microbial Technology*, 35(2-3):126-139.

Linder, M.C. (1985). Nutritional Biochemistry and Metabolisms with Clinical Applications. New York: Elsevier Science publishing company Inc.

Mildner, P., & Ries, B. (Eds.). (2014). Enzyme regulation and mechanism of action: proceedings of the FEBS Special Meeting on Enzymes, Cavtat, Dubrovnik, 1979 (Vol. 60). Elsevier.

Purich, D. L., & Allison, R. D. (2003). The Enzyme Reference: a comprehensive guidebook to enzyme nomenclature, reactions, and methods. Elsevier.

Satyanarayana, U. and Chakrapani, U. (2013). Biochemistry. 4th ed. Kolkata: Elsevier Books and Allied (P) Ltd.

Singh, N.P. (2005). Biochemistry. New Delhi: Akansha Publishing House.

Smith, H. J., & Simons, C. (Eds.). (2004). Enzymes and Their Inhibitors: drug development (Vol. 2). CRC press.

Talwar, G. P., Srivastava, L. M. and Moudgil, K. D. (1989). Textbook of Biochemistry and Human Biology, 2nd edn. New Delhi: Prentice Hall of India Pvt. Ltd.

Voet, D., Voet, J. G. and Pratt, C. W. (2016). Fundamentals of Biochemistry: life at the molecular level. 5th edn. Hoboken, NJ: Wiley.

Webb, E. C. (1992). Enzyme nomenclature 1992. Recommendations of the Nomenclature Committee of the International Union of Biochemistry and Molecular Biology on the Nomenclature and Classification of Enzymes (No. Ed. 6). Academic Press.

http://www.chem.qmul.ac.uk/iubmb/enzyme/rules.html

http://www.srmuniv.ac.in/sites/default/files/files/classification.pdf

http://www.chem.qmul.ac.uk/iubmb/enzyme/

http://www.biotechnologyforums.com/thread-1792.html

https://www.slideshare.net/ArunimaSur/site-directed-mutagenesis

https://alevelnotes.com/Factors-affecting-Enzyme-Activity/146

http://cbm.msoe.edu/markMyweb/ggpmResources/documents/aaskActiveSiteKey.pdf

https://www.slideshare.net/jahir143/enzyme-immobilization-20940809

2
Nucleic Acids

Introduction

The nucleic acids are the molecular depots that are responsible for continuity of life by transfer of hereditary information from one generation to next. The sequence of nitrogenous bases containing programmed instruction determines the structure of every cell constituent and eventually every protein. In 1869, Friedrich Miescher, Swiss chemist, isolated an unknown phosphate-rich substance from pus cells named as nuclein, which was quite different from other organic substances like carbohydrate, protein, and fat. Altmann in 1899 ascertained that the nuclein also had some acidic characteristics and changed its name to 'nucleic acid'. Later, nucleic acid was also isolated from salmon sperm nuclei as DNA-protamine complex. In 1953, James D. Watson and Francis H.C. Crick constructed the model of three-dimensional double helix structure of the DNA molecule which could successfully explain DNA replication.

Definition and composition

The nucleic acid occurs in cell nuclei. These macromolecules are either present as free or restrained to proteins in the form of nucleoproteins. They are high molecular weight biopolymer and composed of monomer nucleotide as repeating unit. Carbon, hydrogen, oxygen, nitrogen, and phosphorus make the elemental composition of the nucleic acids.

There are majorly two nucleic acids, deoxyribonucleic acid (DNA) and ribonucleic acid (RNA).

Both are known to be present entirely in plants and animals. On the contrary, a virus cannot have both types and thus, has either RNA or DNA. The chromatin of the cell nucleus usually shows DNA presence whereas 90% RNA exists in the cell cytoplasm and another 10% in the nucleus. Extra-nuclear DNA also exists and present mostly inside chloroplasts and mitochondria. The nucleic acids upon hydrolysis yield three components: phosphoric acid, a pentose sugar (5-carbon sugar) and a nitrogenous base.

Structure and functions

The nucleic acids are polynucleotides; a long string of molecules possessing a series of almost identical units called nucleotides. Each nucleotide consists of three parts:

a) A nitrogenous base

b) Pentose sugar (Five-carbon sugar)

c) Phosphate group

The typically found nitrogenous bases in nucleic acid are purines and pyrimidines. Adenine (A) and guanine (G) are categorized as purines, while cytosine (C), thymine (T), and uracil (U) are known as pyrimidines. Nitrogenous bases A, G, C, T exist only in DNA, while U exists only in RNA. The 2'-deoxyribose is the pentose sugar present in DNA unlike the ribose sugar i.e., present in RNA. 22 -deoxyribose is distinct in the way that 22 carbon of the furan ring has hydrogen (-H) in place of a hydroxyl group (–OH) (Table 2.1). The nitrogenous base attached with pentose sugar is known as a nucleoside and nucleoside along with phosphate groups forms a nucleotide. Different components of nucleotide are shown below (Fig. 2.1).

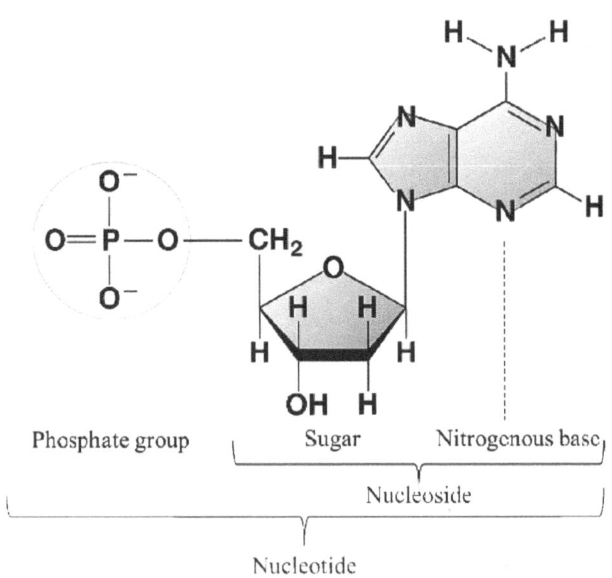

Fig. 2.1 Components of nucleotide

Table 2.1 Composition of DNA and RNA

Components	Ribonucleic acid (RNA)	Deoxyribonucleic acid (DNA)
1. Nitrogenous bases		
a. Purines	Adenine	Adenine
	Guanine	Guanine
b. Pyrimidines	Cytosine	Cytosine
	Uracil	Thymine
2. Pentose Sugar	D-ribose	D-2-deoxy ribose
3. Acid (phosphate group)	Phosphoric acid	Phosphoric acid

Nitrogenous Bases

Nitrogenous bases can be divided into two types: Pyrimidines and Purines.

Pyrimidine

Pyrimidines have a six-membered nitrogen-containing ring structure (Fig. 2.2). Pyrimidines includes thymine (5-methyl-2, 4-dioxypyrimidine), cytosine (2-oxy-4-aminopyrimidine), and uracil (2, 4-dioxypyrimidine) (Fig. 2.3). Cytosine shows its presence in DNA as well as RNA. Uracil is present only in RNA, while thymine primarily present in DNA. Sometimes, tRNA may contain both thymine and uracil.

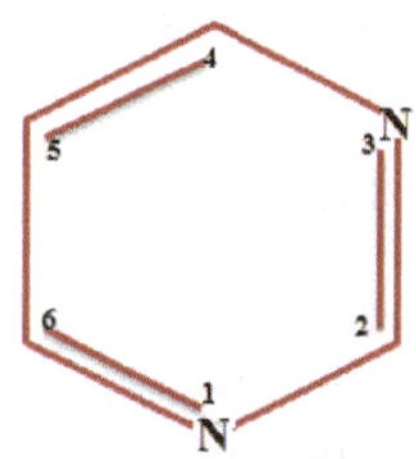

Fig. 2.2 The basic structure of pyrimidine

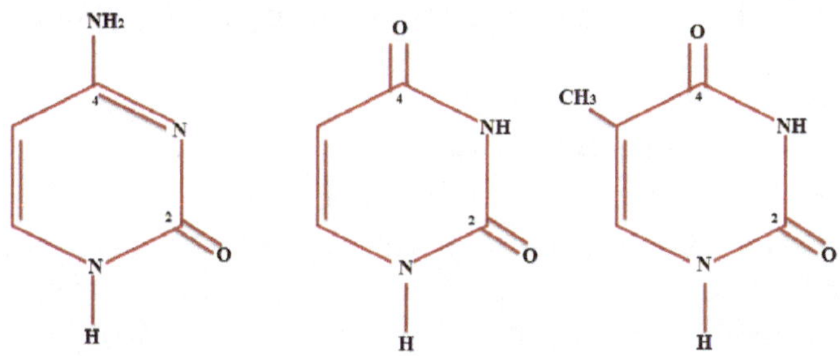

Cytosine (2-oxy-4-aminopyrimidine) Uracil (2,4-dioxypyrimidine) Thymine (5-methyl-2,4-dioxypyrimidine)

Fig. 2.3 The structure of pyrimidine derivatives

Purine

Adenine (6-aminopurine) and guanine (2-amino-6-oxypurine) come under purines and have a basic double ring structure, which comprises of six-membered pyrimidine ring merged with the five-membered imidazole ring as shown in Fig. 2.4 & Fig. 2.5.

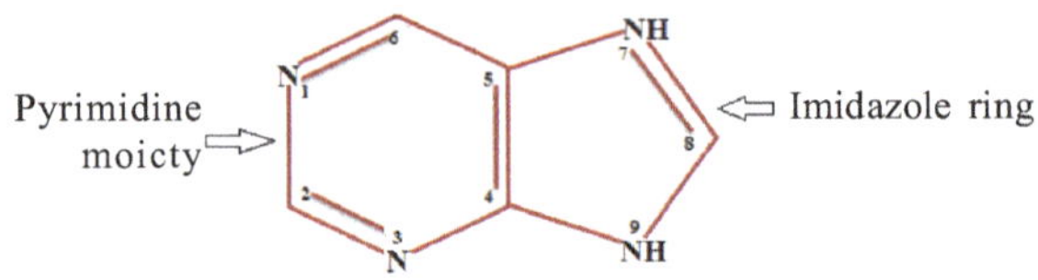

Fig. 2.4 The basic structure of purine

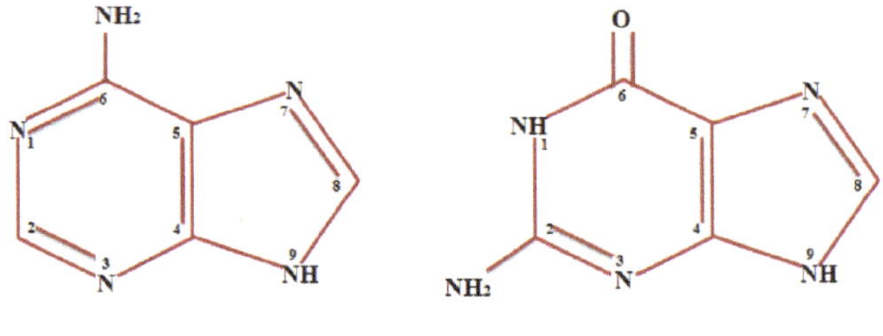

Adenine (6-aminopurine)

Guanine (2-amino-6-oxypurine)

Fig. 2.5 The structure of purine derivatives

Adenine and guanine are likely to be present in both DNA and RNA.

The above mentioned five nitrogenous bases are entitled as major nitrogenous bases due to their large proportion in the total base composition in animal and plant nucleic acids. The other minor nitrogenous bases are also present in nucleic acid and known as modified nitrogenous bases. Other derivatives of purines in plants are also found. For example, coffee contains caffeine (1,3,7-trimethylxanthine) and tea as well as cocoa contains theobromine (3,7-dimethylxanthine). Both caffeine and theobromine possess important medicinal properties.

Tautomerism in Nitrogenous Bases

The compounds having two structural isomeric forms which are mutually inter-convertible and remains in equilibrium with each other are called tautomers and the phenomenon is known as tautomerism. This involves the transfer of a proton (H^+) from α-carbon to carbonyl oxygen as shown in Fig. 2.6.

Tautomerism

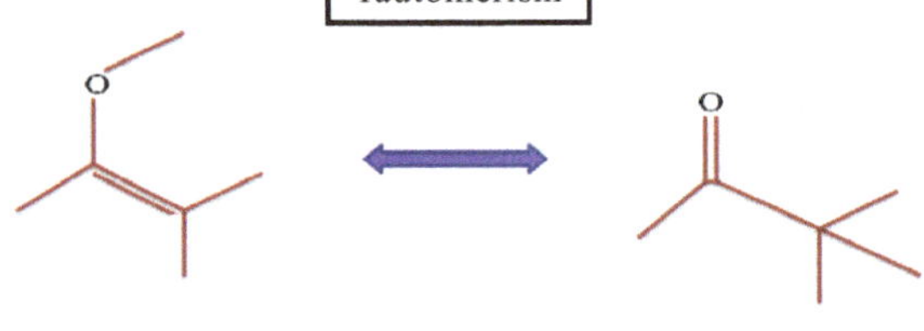

Enrol or Lactim form

Keto or Lactam from

Fig. 2.6 Tautomerism

The carbonyl group (=CO) comprising tautomer form is named as the keto or lactam and the double-bonded carbon linked by hydroxy group (-OH) is named as the enol or lactim form. This type of tautomerism is known as keto-enol or lactam-lactim.

Pentose sugar

DNA and RNA are distinct from each other due to the difference in five-carbon sugar present in them. DNA contains D-2' -deoxyribose, suggesting the name deoxyribose nucleic acid or deoxyribonucleic acid, whereas, RNA contains D-ribose, thus the name ribose nucleic acid or ribonucleic acid (Fig. 2.7). Both D-2'-deoxyribose and D-ribose sugars in nucleic acids are present in the form of furanose in β-configuration. D-ribose is the main sugar while D-2' -deoxyribose is a derivative in which OH group is replaced with H atom at C-2' position. The various carbon atoms of pentose sugar are entitled by a prime, (') sign on numbers, i.e., C-1', C-2', C-3' etc.

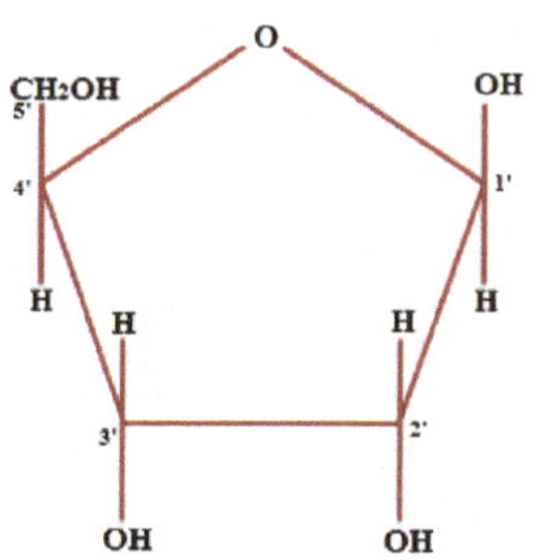

D-ribose
(β-D-ribofuranose)

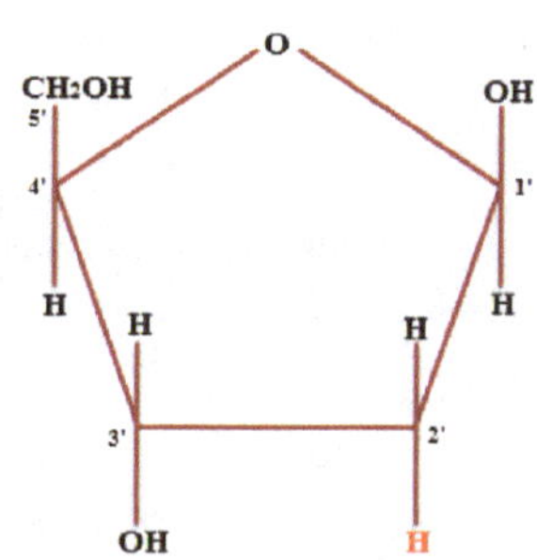

D-2-deoxyribose
(β-D-2-deoxyribofuranose)

Fig. 2.7: The structure of D-ribose and D-2'-deoxyribose

Phosphoric acid

Its molecular formula is H_3PO_4. The structure depicts the presence of three monovalent -OH groups and a divalent oxygen atom attached to central pentavalent phosphorus atom (Fig. 2.8).

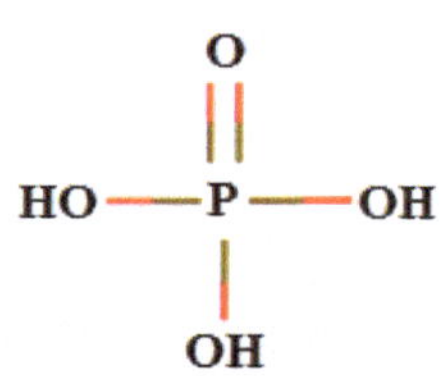

Phosphoric acid

Fig. 2.8 Phosphoric acid

Nucleoside

The nucleosides may have nitrogenous bases such as purines and pyrimidines, which bond with the pentose sugar (ribose or deoxyribose) by a β-glycosidic linkage (Fig. 2.9).

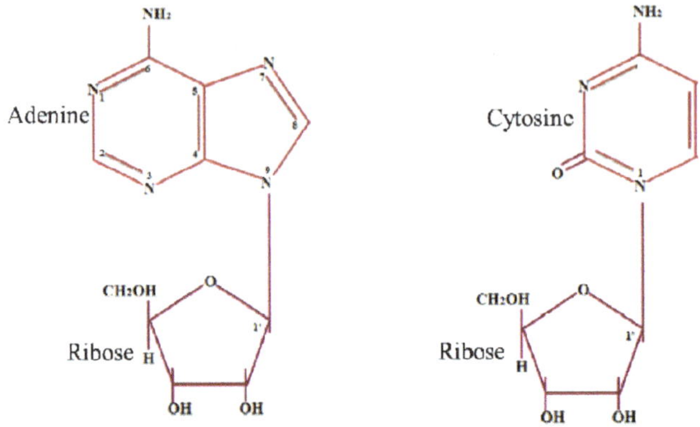

Adenosine Deoxycytidine

Fig. 2.9 Examples of Nucleoside (Adenosine and Deoxycytidine)

The β-glycosidic linkage in formed between the C-1 of sugar and the nitrogen atom of purines (at 9 position) or pyrimidines (at 1 position). The nucleosides are named after peculiar purine or pyrimidine present. Nucleosides embracing ribose are termed as ribonucleosides, whereas those containing deoxyribose are termed as deoxyribonucleosides (Table 2.2).

Table: 2.2 Nucleosides and their nomenclature

Base	Sugar	Nucleoside	Trivial name
Adenine	Ribose	Adenine ribonucleoside	Adenosine
Guanine		Guanine ribonucleoside	Guanosine
Cytosine		Cytosine ribonucleoside	Cytidine
Thymine		Thymine ribonucleoside	Thymidine
Uracil		Uracil ribonucleoside	Uridine
Adenine	Deoxyribose	Adenine deoxyribonucleoside	Deoxyadenosine
Guanine		Guanine deoxyribonucleoside	Deoxyguanosine
Cytosine		Cytosine deoxyribonucleoside	Deoxycytidine
Thymine		Thymine deoxyribonucleoside	Thymidine (or Deoxythymidine)*
Uracil		Uracil deoxyribonucleoside	Deoxyuridine

Note: The trivial names for the pyrimidine (nucleosides) end with the suffix -dine, while purine (nucleosides) end with the suffix -sine.
*Since Thymine is primarily present in DNA, thus deoxyribonucleoside is termed as thymidine instead of deoxythymidine. Moreover, it also appears in transfer RNA (tRNA).

Nucleotides

Nucleoside along with a phosphoric acid molecule is termed as a nucleotide. Thus, every nucleotide comprises three units: a nitrogenous base, a pentose sugar, and a phosphate group (one or more). The pentose sugar molecule acquires a central position in the nucleotide structure, the nitrogenous base coupled to one of its carbons and the phosphate group (or groups) coupled with another. It primarily formed in a nucleic acid as a subunit or occurs in the free form.

Since C1' and C4' of ribose moiety of ribonucleoside are engaged in the furanose ring formation, hence the phosphorylation is viable only at three carbons positions viz. C2', C3', C5'. On the contrary, phosphorylation can occur at C3', C5' in the deoxyribose moiety of a 2'-deoxyribonucleoside, as again C1' and C4' engaged in the furanose ring formation and C2' do not have a hydroxyl group.

Nucleotides may have one, two or three phosphate groups bonded covalently to pentose sugar at 5'-C. Starting from the pentose group, the three phosphate groups are normally labeled as α, β, and γ. These are attributed as nucleoside mono-, di- and triphosphates and shortly abbreviated as NMPs, NDPs, and NTPs, respectively (Fig. 2.10).

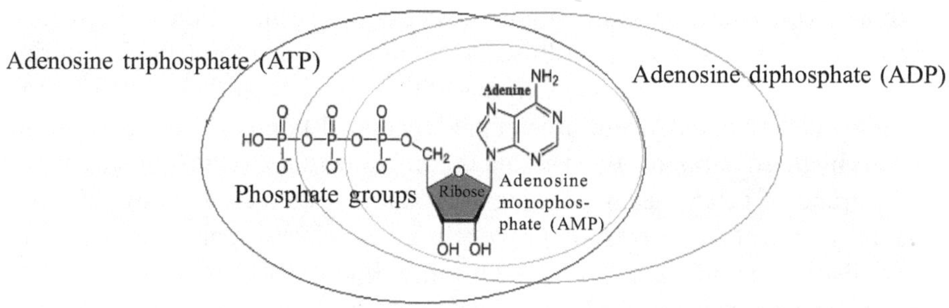

Nucleotide: The phosphate ester and phosphoric acid anhydride bond of ATP

Fig. 2.10 The structure of Nucleotide

Polynucleotide chains

DNA and RNA are polymers and are made up of monomers known as nucleotides. When these monomers are joined with each other, it results in the formation of a polynucleotide chain (*poly* means "many"). Whenever a nucleotide joins the enlarging polynucleotide chain of DNA or RNA, it releases two phosphate groups. Thus, every nucleotide has only one phosphate group in polynucleotide chain. The chain has directionality i.e., the two ends are distinct from each other. At the beginning or 5' end of the chain, first nucleotide's 5' phosphate group in the polynucleotide chain protrudes. Similarly, at the other end of the 3' end, the lagging nucleotide joins and sticks out the 3' hydroxyl group. Thus, the DNA or RNA strands sequence is usually penned in the 5' to 3' direction, meaning that 5' end of nucleotide comes first (starting point) and 3' end of nucleotide comes at the end (last point) (Fig. 2.12).

When new nucleotides sum up in a strand of DNA or RNA, the 5' phosphate of newer nucleotide gets tied to the hydroxyl group at the 3' end of the chain and strand usually grows in direction of 3' end. In a polynucleotide chain, every pentose sugar sticks to its neighbor's phosphate group by bond known as phosphodiester linkage (Fig. 2.11).

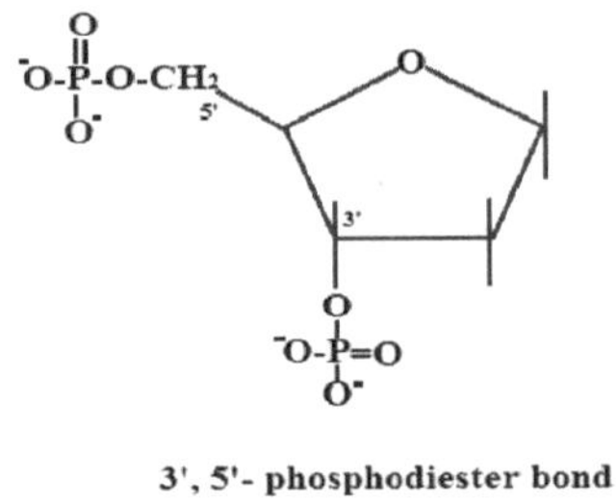

Fig. 2.11: 3', 5'-phosphodiester bond

Structure of DNA

DNA is a cord of deoxyribonucleotides. It is usually comprised of two polynucleotide chains going parallel but in the reversed direction and intertwined to form the double helical structure of DNA. The constituent units i.e., a nitrogenous base, deoxyribose sugar, and a phosphate group are coupled by 3', 5' -phosphodiester bonds. These phosphodiester linkages join sugar moieties and act as polynucleotide backbone, which performs a structural role in DNA. The glycosidic linkage bonds the nitrogenous bases and deoxyribose. N1 of the pyrimidines or N9 of the purines are linked with deoxyribose. Two strands of DNA are held together by hydrogen bonding: adenine and thymine are coupled by two H-bonds, while, guanine and cytosine by three H-bonds. Each strand is complementary to the opposite strand. If one strand of DNA has an adenine at a position, then another strand would have a thymine at the corresponding position. Similarly, guanine and cytosine would also be complementary to each other. Each DNA strand has a free 5' phosphate or 5' hydroxyl at one end and a free 3' phosphate or 3' hydroxyl at the other end. DNA sequences are read from 5' to 3' end direction (on the left) as demonstrated in Fig. 2.12, generally with a 5' phosphate and a 3' hydroxyl end. The orientation of the DNA strands is said to be anti-parallel.

Because of the DNA's double-helical structure, two grooves are formed i.e., major groove and minor groove, that are not equal in size to each other. In a DNA structure, because of the stacking of amassed bases in DNA structure, two sugars extrude from the base pairs and narrow-angle of 120° or wide angle of 240° is formed. The minor groove is formed due to the narrow-angle at one edge of the base pairs and major groove due to wide angle on the other edge.

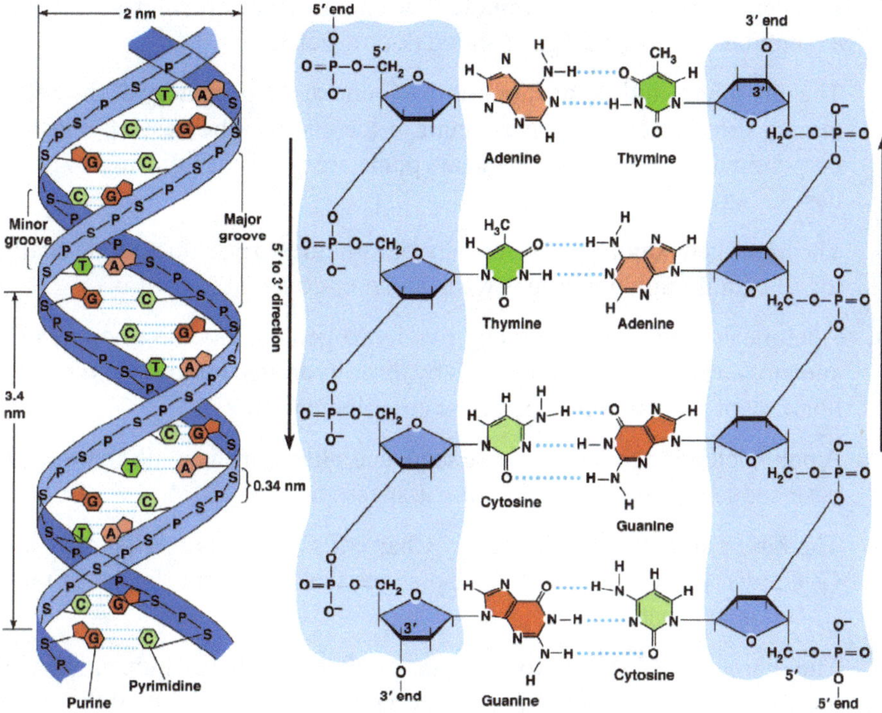

Fig. 2.12: General structure of B-DNA

Chargaff's rule of DNA composition

In 1950, Erwin Chargaff discovered the pattern of DNA strands and gave two major findings:

(a) In a double-stranded DNA, the numbers of guanine units are comparable (equal) with the number of cytosine units; and the numbers of adenine units are comparable with the number of thymine units; known as first parity rule.

(b) The composition of DNA diverges from one species to another; known as second parity rules. Chargaff's findings laid down the basis of Watson and Crick's double-helix DNA model in 1953.

Conformations of DNA double helix

In 1953, James Watson and Francis Crick constructed the three-dimensional double helical model of B-DNA on the basis of X-ray diffraction data. The noticeable features of this model are:

1. DNA encompasses two polynucleotide chains which are intertwined over a common axis to give right-handed double helix.

2. The phosphate and sugar subunits are found at the periphery of the helix, though the nitrogenous bases occur at the centre. The bases planes are perpendicular to the axis and sugars plane are nearly at right angles to the bases.

3. The hydrogen bonding between base pairs held two polynucleotide chains together, although bonding is weak and non-covalent in nature.

4. Adenine (A) invariably pairs by two hydrogen bonds with thymine (T); and guanine (G) with cytosine (C) by three hydrogen bonds. This specific bonding of the bases is called base complementarity.

5. A polynucleotide chain has a base sequence, although changing and specific base sequences transmit the genetic message.

6. The base content of DNA comply Chargaff s rules, conferring A=T and G=C; also (A+C) = (G+T). It also states that proportion of (A+T) and (G+C) is invariable for a species.

7. The diameter of DNA helix is 20nm or 20 Å and the neighboring bases are spaced by 0.34 nm or by 3.4 Å along the axis. The length of the entire turn of the helix is 3.4 nm or 34 Å *i.e.*, 10 base pair per turn. In other words, each turn in helix contains 10 nucleotide residues.

8. The two helices of DNA are intertwined over a common axis and the glycosidic bonding with base pairs are not diametrically contrary to each one, thus, two grooves are formed, a major groove with 12 Å width, 8.5 Å depth and a minor groove with 6 Å width, 7.5 Å depth. In these grooves, peculiar proteins interact with DNA molecules. The unwinding of DNA results in the abstraction of two helices or coils that are intertwined about the same axis and termed as *plectonemic coils*.

9. These two helices or coils run antiparallel to each other, means in opposite direction. As the base pairing is *complementary* to each other, hence are not alike.

10. These DNA strands have reverse polarity, means the two polynucleotide chains sequence of the units is opposite to each other. If the one strand ascends, runs in the 5'→3' direction then, other one descends and runs in the 3'→5' direction.

Alternative DNA structures

DNA prevails in many possible conformations such as A-DNA, C-DNA, D-DNA, Z-DNA and B-DNA. B-DNA is the most common form found in organisms.

A-DNA: It is a right-handed double helical form. This variant appears due to dehydration of B-DNA, primarily by reduction of relative humidity from 92 to 75% plus the presence of Na^+, K^+, and Cs^+ ions. Its phosphate groups have a lesser affinity for water molecules as compared to B-DNA. It comprises of 11 base pairs per helical turn and has 2.3 Å of vertical rise per base pair in place of the 3.4 Å rise. The helix pitch means the total distance moved by helix is (2.3×11) or 25.30° per turn instead of (3.4 × 10.4) or 35.36° in B-DNA. Apparently, the revolution per base pair is 360/11 or 32.72° then the 360/10.4 or 34.61° found in B-DNA. The A-DNA helices are wider and shorter relative to B-DNA. Its helical diameter is 25.5 Å. In this, C3' is out of a plane in sugar furanose ring and made by another four carbon atoms, that causes a puckering or wrinkling in the ribose sugar. Furthermore, the minor groove does not exist in A-DNA practically, while B-DNA has a quite deep minor groove.

C-DNA: This pattern is made at a relative humidity of 66% in the Li^+ ion presence. This is also right-handed and the axial rise is 3.32 Å per base pair. The base pair per turn is 9.33 and thus the helix pitch is 3.32 × 9.33 Å or 30.97 Å. The pitch diameter is 19 Å, fewer than that of other forms of DNA. Its revolution per base pair is 360/9.33 or 38.58°. The base pairs are inclined to 7.8° in C-helix.

D-DNA: This variant is exceptionally uncommon and has only 8 base pairs per helical turn. This usually appears in some DNA molecules lacking guanine. Theaxial rise is 3.03 Å per base pair and is seen to be inclined at 16.7° from the helical axis.

Z-DNA: Z-DNA is characterized by a left-handed double helix. The recurring dinucleotide unit amid sugar phosphates in the DNA backbone gives a *zig-zag* structure; hence, termed as Z-DNA. Another exceptional feature is that in it the adjoining sugar residues have alternating orientation, whereas, in B-DNA, the orientation is same amid the adjoining sugar residues. Hence the recurring unit is a mononucleotide. It has only one deep helical groove. It is composed of 12 base pairs per helical turn and the axial rise is 3.8 Å per base pair. The angle of inclination of helical axis is 9°. As in one helix, 12 base pairs are accommodated unlike other variants; the angle of spin per recurring unit is 60°. The length of one complete helix is 45.60 Å contrary to 35.36 Å of B-DNA. The bases in Z-DNA have more length to escalate and the angle of inclination is 60°, thus near to the axis; hence, its diameter is 18.4 Å rather than 23.7 Å which formed in B-DNA. The other characteristics are shown in Table 2.3.

Table 2.3 Corresponding structural properties of A, B, C and Z DNAs

Characteristics	A DNA	B DNA	C DNA	Z DNA
Conditions	75% relative humidity; Na^+, K^+, Cs^+ ions	92% relative humidity; low ion strength	60% relative humidity; Li^+ ions	Very high salt concentration
Shape	Broadest	Intermediate	Narrow	Narrowest
Helix sense	Right-handed	Right-handed	Right-handed	Left-handed
Helix diameter	25.5 A°	23.7 A°	19.0 A°	18.4 A°
Rise per base pair 'h'	2.3 A°	3.4 A°	3.32 A°	3.8 A°
Base pair per turn of helix 'n'	11	10.4	9.33	12
Helix pitch 'h x n'	25.30 A°	35.36 A°	30.97 A°	45.60 A°
Revolution per base pair	+ 32.72°	+ 34.61°	+ 38.58°	+ 60°
Base pair tilt	19°	1°	7.8°	9°
Glycosidic bond syn for A,G	anti-conformation	anti-conformation	-	anti for C,T;
Major groove	Narrow and very deep	Wide and quite deep	-	Flat
Minor groove	Very broad and shallow	Narrow and quite deep	-	Very narrow and deep

Other related structures of DNA

Bent DNA: In this type of DNA, frequent bends are found in the structure due to the presence of four or more adenine bases that repeat sequentially in one of the two chains. The bend of around 18° appears due to the existence of six adenines in a row. Bending may have importance in DNA structure for confining some proteins.

Palindromic DNA: The term palindromic DNA implies to the domain in DNA, when the base sequence of two strands shows inverted repetitions. Such nucleic acid sequences are self-complementary and when only a single stand is available, likely to make hairpin; and cruciform structures when both strands of a DNA present. The term mirror repeat is used for DNA sequence when the upturned sequence recurs within each exclusive strand. Mirror repeats would not make hairpin or cruciform structures due to lack of complementary sequence within each strand. A sequence of these types appears in large DNA molecule which includes either few or thousands of base pairs.

H-DNA: H-DNA generally originates in polypyrimidine or polypurine sequences that contains mirror repeat. A long stretched sequence is made with two nitrogenous bases alternating each time. H-DNA's striking feature is the pairing and twisting of three strands, thus a triple helical structure. It is produced instinctively by two chains comprising only pyrimidines and the third one comprises of purines.

Denaturation of DNA

Denaturation is the loss of biological activity due to the unwinding of two polynucleotide chains, where the two strands open up and separate as a single stand. During the process, the DNA molecule splits up into two strands because of weak and non-covalent hydrogen bonding present between the bases. The changeover from innate (native) to a denatured form occurs abruptly and is favored in presence of urea and formamide, which makes the purine and pyrimidine groups more soluble in water. The denaturation of the double helical form of DNA can be done by following denaturing agents:

Denaturation by temperature: The DNA double helix can be altered or denatured by the addition of alkali, acid or by heat. The *denaturation temperature* or *melting temperature* (T_m) is the point at which half of the helical structure is wiped out. This opening of the two helical strands of the double helix is known as *melting*. Denaturation causes the following changes in properties of DNA. These are:

a) ***Increase in the absorption of ultraviolet light:*** Each base in nucleic acids absorbs ultraviolet light and is characterized by a maximum absorption of UV light at 260 nm. When the native DNA denatures, a marked increase can be observed in the absorption of UV light. This effect is called as hyperchromicity.

b) ***Decrease in specific optical rotation:*** The double helical DNA exhibits a strong positive specific rotation which subsequently decreases with denaturation.

c) ***Decrease in viscosity:*** Viscosity reduces as DNA denatures due to the splitting of hydrogen bonds. The solutions of native DNA are highly viscous because of the relatively rigid double helical structural forms plus long rod-like structures.

Denaturation by chemical agents: The certain chemicals like urea and formamide disrupt the H-bonds and denature the DNA double helix model. These chemical reagents work by enhancing the water solubility of the purine and pyrimidine groups. Melting temperature (T_m) value is reduced by urea addition in DNA. DNA completely denatures in 95% formamide at ambient condition.

Effect of pH on denaturation: Acidic and alkaline pH conditions cause ionic changes in the substituent's of the purine and pyrimidine bases and thereby denature DNA. At acidic pH of 2 to 3, amino groups accept protons and disrupted the DNA helix structure. Likewise, at alkaline pH, the enolic hydroxyl group ionizes and thereby prevents the hydrogen bonding between keto and amino groups.

Renaturation of DNA: When a DNA denatured solution is cooled at slow rate to an ambient temperature, renaturation of some part of DNA occurs. The maximum reversibility of 50-60% is attainable by slow cooling. In this process, the denatured DNA solution is held at a temperature of about 25° C and a concentration of 0.4 M Na^+ or more for several hours. Quick cooling would not renature denatured DNA; but if the cooled solution is heated again and cooled slowly then it induces renaturation. When the complementary bases reunite by hydrogen bonding, this causes the restoration of the double helical structure.

Organization of DNA in the cells

Cells contain a nucleus encircled by a nuclear membrane in eukaryotic cells, and a nuclear region in the prokaryotic cells. In a eukaryotic cell, the nucleus is present in a complex form bound with a thread-like structure called "chromatin". Chromatin is a complex of macromolecules like DNA, proteins, and RNA. Its primary function is to wrap DNA into a very compact form inside the cell nucleus. DNA wraps around the histone proteins to form condensed nucleosomes. Nucleosomes or bead on a string is the basic unit of DNA packaging in which over 200 base pairs of DNA are coated around histone proteins. Thus, huge long pleated DNA molecule coupled with histones is packed into a much compact form.

Structure of RNA

RNA (Ribonucleic acid) is a long, unbranched macromolecule comprised of nucleotides joined by 3' $\rightarrow$ 5' phosphodiester bonds. Unlike DNA, the RNA molecules possess numerous specific differences. The RNA has ribose sugar moiety *rather* than 2'-deoxyribose of DNA to which the phosphate and nitrogenous bases are linked. RNA is composed of the uracil (U) instead of thymine. Uracil is bonded with adenine (A) by two H-bonds. The inherent RNA molecule is single-stranded, contrary to characteristic double-stranded helical form. However, the single strand may acquire a hairpin structure and thus, a double-stranded pattern can be achieved where, A pairs with U; and G pairs with C. The G-U pair can also be formed but it is comparatively weak. Since RNA is single-stranded, its adenine content is not equal to its uracil content; as well as guanine content is not equal to cytosine content.

Functions of RNA

1. RNA is not the hereditary material (except viruses) as it is not capable of replication. It has been found to be genetic material in TMV (Tobacco mosaic virus), ϕ β bacteriophage etc. DNA is major hereditary material in most of the organisms. Rather, RNA has intermediary functions; the mRNA passes the information from the gene for the protein synthesis.
2. It is a regulatory molecule and takes parts in the translation process.
3. It acts as an adaptor for protein synthesis. The tRNA transfers the codons on the mRNA in order to synthesize proteins.
4. It has a structural role in the form of rRNA.
5. RNA enzymes are known as **ribozymes**, and they may have many similarities with a classical enzyme due to the presence of an active site, a substrate binding site and a binding site for a cofactor.

The three usual kinds of RNAs are found in all organisms: *ribosomal, messenger and transfer RNAs.*

1. *Ribosomal RNA (rRNA)*

Ribosomal RNA is extremely ample form among all other types of RNA. It is the most stable form resides in ribosomes. It accounts for 80% of the total RNA of a cell and depicts about 40-60% of the total weight of ribosomes. Ribosomes have a large subunit called as the 50S and a small subunit called a 30S, each of which has its own rRNA. They are present as small rRNAs and large rRNAs units; and signify their existence in the small and large subunits of the ribosome. These rRNAs join with proteins inside cytoplasm and forms ribosomes, which then serve as the sites for protein synthesis. These ribosomes transit along with mRNA molecule during the translation process and assist in the organization of amino acids into a polypeptide chain. The tRNAs transfers the triplet codon to this assembly which is essential for protein synthesis.

2. *Messenger RNA (mRNA)*

It is the most heterogeneous in size and stable among all the types of RNAs. It accounts for 5% of total RNA of a cell and has large molecular weight. The hereditary message copied from the DNA during transcription process is carried by mRNAs in the form of triplets of nucleotides termed as codons, to the ribosomes (site of protein synthesis). Each codon explicitly codes for a particular amino acid. Moreover, one amino acid can be coded by several different codons. There are possible 64 codons consisting of three stop codons representing 20 amino acids.

3. *Transfer RNA (t-RNA)*

Transfer-RNA is the smallest in size and accounts for about 15% of total RNA in the cell. The most substantial role of t-RNA is to serve as a carrier for amino acids to specific sites (i.e., a template of protein synthesis on ribosomes) during the translation process. There are 20 amino acids known, and each amino acid is coded and has a specific t-RNA that transfers and binds these amino acids into an enlarging polypeptide chain for protein synthesis. Thus, t-RNA acts as adapters during translation of the mRNAs. The one amino acid may have several t-RNAs, sometimes up to 4 or 5. They are capable of confining the same amino acid with altering nucleotide sequence; either recognize the alike codon by the alike anticodon or have different anti-codons. Thus, t-RNA organizes the amino acid for protein synthesis in response to multiple codons explicitly for the same amino acid.

A cloverleaf structure of t-RNA is shown in Fig. 2.13. Apart from the four major bases, t-RNAs also contain some unusual bases primarily results from methylation of the major bases, e.g. methyl guanine and methylcytosine.

The structure of t-RNA consists of the following feature:

1. *Amino acid arm*

Amino acid binds to this arm and thus, it is named as amino acid arm/CCA arm or amino acid acceptor arm. There are 7 base pairs between 5' and 3' ends of tRNA amino acid arm. A sequence of 5' -CCA-3' is added at 3' end.

2. *D-arm*

D-arm comprised of modified base dihydrouracil i.e., 3 to 4 base pair stem and a loop known to be D-loop or DHU-loop.

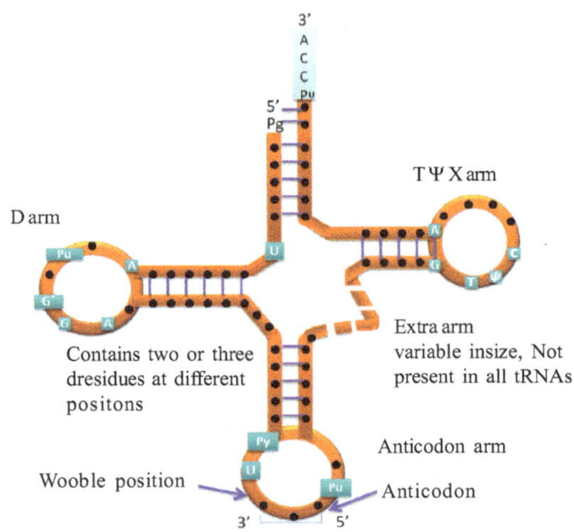

Fig. 2.13: The structure of tRNA

3. Anticodon arm

It lies contrary to the acceptor's arm. Anticodons complements to the codon of mRNA located in a loop and have 5 base pairs.

4. An extra arm

An extra arm contains over 3-21 bases and two of its types is known as a small extra arm with 3-5 bases and a large arm with 13-21 bases, respectively.

5. T-arm or Tψ C arm

It encloses a modified base (pseudouridine ψ) and a 5 base pair stem with a loop. Besides the four major bases, around 50 distinct types of modified bases also exist. It contains ribothymidine base not existent in RNA. Some modified bases are pseudouridine ψ, dihydrouridine, and inosine etc.

Sample questions

1. Define nucleic acid and briefly discuss its biological significance?
2. Why is uracil not present in DNA? Suggest the possible reason.
3. What do you mean by tautomers and tautomerism? Give a brief along with example.
4. What are the components of a nucleotide? How the components are linked together in a nucleotide? Give the structure also.
5. Draw the well-labeled structure of ATP?
6. Write down the salient features of Watson and Crick model of double-stranded DNA.
7. Compare the different forms of DNA and their occurrence in nature?
8. Discuss the Chargaff's rule of DNA composition.
9. What do you understand by hyperchromicity of denaturation in DNA?
10. Differentiate between:
    a. Nucleoside and nucleotide
    b. Ribonucleotide and Deoxyribonucleotide
    c. DNA and RNA
    d. Purine and pyrimidine
    e. Denaturation and renaturation
11. Give a brief account of different types of RNA. Discuss the functions of each.

Suggested Readings

Adams, R.L.P. Knowler, J. T. and Leader, D. P. (1992). *The biochemistry of nucleic acids*. 11th edn. London: Chapman and Hall Ltd.

Berg, J. M., Tymoczko, J. L., Gatto, G. J. and Stryer, L. (2015). *Biochemistry*. 8th edn. New York: W.H. Freeman & Company.

Berk, L. H. et. al., (2003). *Molecular Cell Biology*. 5th edn. New York: W. H. Freeman & Company.

Cox, M. M., and Nelson, D. L. (2008). *Lehninger Principles of Biochemistry*. 5th edn. New York: Worth Publishers.

Johnson, A. B. et al., (2002). *Molecular Biology of the Cell*. 4th edn. New York: Garland Publishing, Inc.

Satyanarayana, U. and Chakrapani, U. (2013). *Biochemistry*. 4th ed. Kolkata: Elsevier Books and Allied (P) Ltd.

Talwar, G. P., Srivastava, L. M. and Moudgil, K. D. (1989). *Textbook of biochemistry and human biology*. 2nd edn. New Delhi: Prentice Hall of India Pvt. Ltd.

Voet, D., Voet, J. G. and Pratt, C. W. (2016). *Fundamentals of biochemistry: life at the molecular level*. 5th edn. Hoboken, NJ: Wiley.

Wilson, K. and Walker, J. eds. (2005). *Principles and Techniques of Biochemistry and Molecular Biology*. 6th edn. New York: Cambridge University Press.

3

Vitamins

Introduction

The basic components of diet include carbohydrate, protein, fat, vitamins and minerals. Vitamins are the essential nutrients which are not formed in our body and required in minute amount from the diet. Vitamins are crucial for the optimal functions of several enzymes; thus important for optimal growth and development. Vitamins perform various functions that help in regulation of metabolism, prevent chronic diseases, maintain natural appetite, mental health and immunity.

In 1912, Polish biochemist Kazimierz Funk coined the term 'vitamine';a substance that was present in rice polishing and foods that cured polyneuritis in birds and beriberi in humans. The name is from *'vit'* for *vital* nutrient and *'amine'* for its similarity as an *amine*. Vital signify that these factors are indispensable for life. Earlier they are identified and thought to contain amino groups, thus the word amine was used, which was true for thiamine but later work revealed that most of the vitamins do not have amino groups. Thus the word was shortened to *vitamin* in English. It should be noted that the term vitamin does not denote to some othern utrients such as essential amino acids, essential fatty acids, dietary minerals or other nutrients that contribute towards good health.

Vitamins may be defined as organic compound needed in minute amount through diet to perform specific biological functions for normal growth and health of the organisms.Vitamins should have following characteristics:

a) Must be obtained through diet as the body cannot synthesize wholly or in adequate amounts.

b) Must be taken regularly as they are stored in limited quantities and get lost gradually.

There are vitamins like substitutes or compounds that fail to meet all the criteria necessary to be classified as vitamins. These compounds may have some properties similar to vitamin or may appear in huge amounts. There are some compounds which can be synthesized in appropriate amounts to meet its needs if precursors are present; some such substances are bioflavonoids, inositol, choline, para-aminobenzoic acid (PABA), carnitine and lipoic acid.

Classification of vitamins

In total, there are thirteen vitamins which are universally recognized. Vitamins are classified into two classes on the basis of their solubility as demonstrated in Fig 3.1.

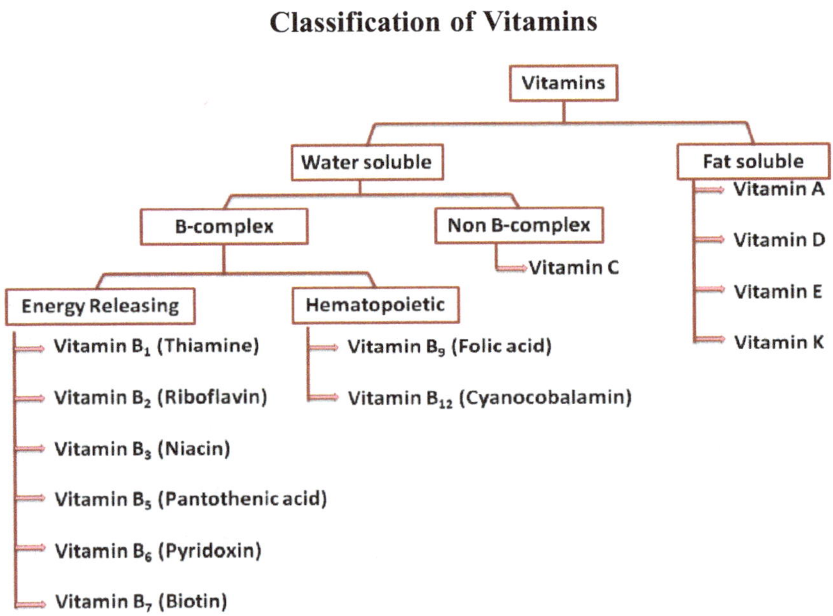

Fig. 3.1: Classification of vitamins

Fat-soluble vitamins

The fat-soluble vitamins, A, D, E, and K are soluble in lipid. They perform certain biochemical functions and excess of these could be reserved in the liver for long periods of time. Only a small amount of these vitamins is required on daily basis. Excess consumption of fat soluble vitamins may have a toxic effect, compared to water-soluble vitamins as being fat-soluble they does not excrete out of the body through urine and their excess accumulation may cause toxicity (hypervitaminosis). For RDA of different vitamins, please refer to Table 10.2 in Annexure.

Vitamin A

It is a fat-soluble vitamin, also called as retinol. Other names are an anti-infection or antixerophthalmic vitamin.

Chemistry of Vitamin A

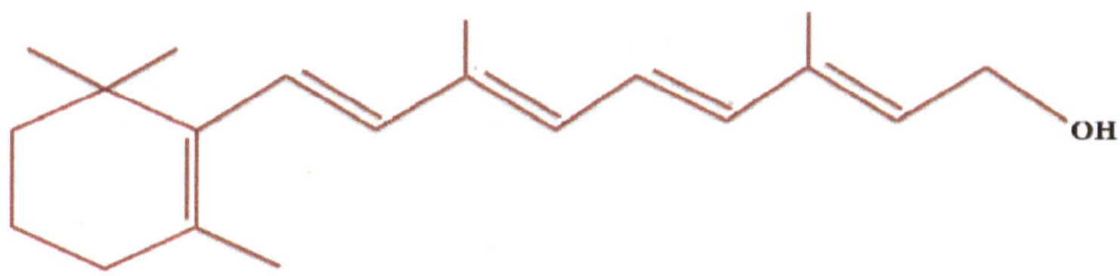

Fig 3.2: Structure of Vitamin A (Retinol)

Vitamin A (retinol) is a generic term for a large number of related unsaturated organic compounds which covers retinol, retinal, retinoic acid, various provitamin A carotenoids and related compounds. Retinol, dehydro-retinol, retinal, retinoic acid, retinyl esters and neovitaminA are termed as retinoids. The two prime forms of vitamin A can be retrieved from food; retinol from animal and carotenoids as provitamin A from plant sources. All these forms of vitamin A are vital for proper growth, reproduction, vision and differentiation of epithelial cells. Retinol and carotenes, both are fat soluble. Vitamin A structural features comprise a β-ionone ring to which an isoprenoid chain is linked, called a retinyl group (Fig. 3.2). Both of these structural features are imperative for vitamin A activity.The carotenes α-carotene, β-carotene, γ-carotene; and the xanthophyll β-cryptoxanthin have a certain level of vitamin activity; other carotenes do not possess this. β-carotene act as a pro-vitamin (precursor for vitamin A) that can be transformed to retinol in intestinal mucosa and is characterized by two connected retinyl groups, thus, notably contribute to vitamin A activity. In contrary, single retinyl group gives α-carotene and γ-carotene some vitamin activity. A provitamin, β-cryptoxanthin (closely related to β-carotene) also changed to retinol in humans. The vitamin activity depends on enzyme β-carotene15,15'-dioxygenase present in herbivores and omnivores animals, which cleaves beta-carotene in the intestinal mucosa and converts it to retinol (Fig. 3.3).

Carnivores are poor converter and pure carnivores cannot convert at all because of lack of β-carotene 15,15'-dioxygenase, thus none of the carotenoids are being converted into vitamin A.

In 1967, joint FAO/WHO Expert Group recommended the concept of the retinol equivalent (RE) to assert the vitamin A activity and enact relationships among vitamin A, these are:

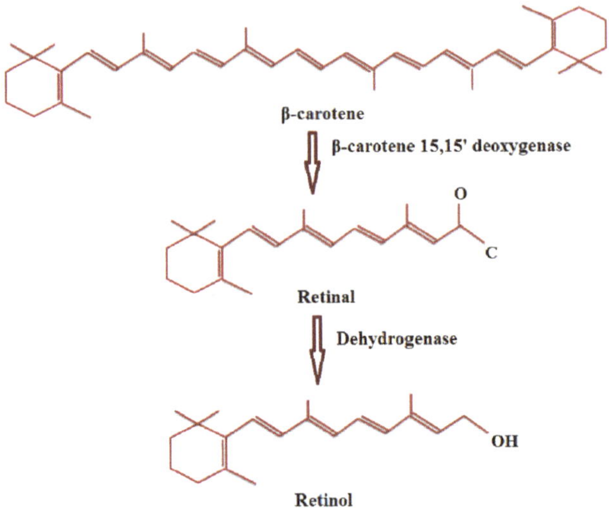

Fig 3.3: Vitamin A activity

1 μg retinol = 1 RE

1 μg β-carotene = 0.167 μg RE

1 μg other pro-vitamin A carotenoids = 0.084 μg RE

These equivalencies are attained from balance studies which exhibit the inefficient absorption of carotenoids and their bioconversion into vitamin A.

Acquired vitamin A is reserved in the liver as retinyl palmitate. Zinc is requisite for retinyl palmitate activity as free retinol. Free retinol is highly active, but toxic, and is therefore carried in the bloodstream with retinol-binding protein (RBP). Moreover, by this complex association vitamin A is solubilized and secured from excretion. In lack of vitamin A, RBP formation in liver diminishes.

Biochemical functions

Vitamin A primarily works by keeping up the vision, notably low light; maintaining growth and integrity of the cells. It is carried to ocular tissue and retina of the eye by intracellular binding with protein molecule (scotopsin, an opsin) plus covalently bound cofactor (retinal); thus, make rhodopsin, a visual pigment pivotal for dim-light vision. Largely, the transformation from all-trans retinol into retinaldehyde in rod cells, isomerization to the 11-cis form and binding to opsin lead to the occurrence of rhodopsin. Through series of photochemical reactions, resulting in a change in rhodopsin and thus enable low light vision. The pace with which rhodopsin is reinvigorated relates to the usable retinol. When there is the scarce amount of retinol in the body, altering rhodopsin synthesis and may result in night blindness. However, deficiency of other nutrients like protein & zinc, which are vital for rhodopsin regeneration or some genetic diseases like retinitis pigmentosa may also be responsible for night blindness (Fig. 3.4).

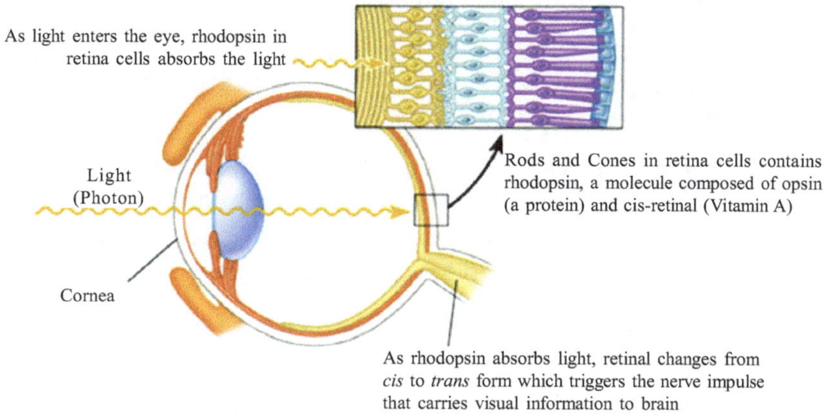

Fig. 3.4: Visual cycle of eye

Vitamin A plays an important role in maintaining the integrity of cells, normal functioning of epithelial & glandular tissues and the tissues that line respiratory, intestinal and urinary tracts. It is anti-infective vitamin; hence maintains the integrity of epithelial tissues by refraining incursion of pathogens. β-carotene has anti-oxidant property and it scavenges free radicals; thus, reduces the occurrence of lung, breast, oral, esophageal and bladder cancers.

Dietary sources

Vitamin A can be acquired as:

(a) ***Preformed vitamin A*** from animal sources like egg, meat, fish liver oil, whole milk, and milk products. Cow milk contains about 46μg of vitamin A equivalent.

(b) ***Provitamin A carotenoids*** from plant sources like green leafy vegetables (spinach, amaranth) and yellow-colored vegetables (carrots, pumpkin) and non-citrus yellow-orange colored fruits (papaya, mango, apricot). Red palm oil and other indigenous plants like palm fruit named are also rich repositories of the vitamin.

(c) ***Fortification with preformed vitamin A*** in whole milk and dairy, processed foods etc.

Deficiency manifestations

The deficiency of vitamin A manifests only when body stores are not sufficient to meet the requirements. The manifestations are primarily seen in eye and skin. Some of the most common manifestations are as follows:

a) ***Night blindness:*** This is the earliest characteristic of vitamin A deficiency, resulting in defective dark adaptation. Such patients are not able to read or drive a car in poor light.

b) ***Xerophthalmia:*** The conjunctiva in the eye becomes dry, thick and wrinkled. It gets keratinized and mislays its natural transparency by acquiring a hazy appearance. When it reaches cornea and cause keratinization of corneal epithelium, becomes glazy and lusterless.

c) ***Bitotsspot:*** They are greyish-white triangular or irregular plaques of keratin, firmly stuck to the conjunctiva.

d) ***Keratomalacia:*** It is characterized by softening of the cornea and is succeeded by the long persistence of xeropthalmia. Further perforation of cornea and prolepses of iris may cause total blindness.

e) ***Follicular hyperkeratosis or phrynoderma:*** It is the result of hyperkeratinization of the epithelium, lining, and the follicles. Skin becomes rough; papules are visible at the site of pilosebaceous follicles.

f) ***Other manifestation:*** The failure in skeleton growth may outdo growth retardation due to a defect in the synthesis of chondroitin sulfate. It may downturn protein synthesis, lowered glycoprotein content of a cell, thus reduced immune competency.

Vitamin D

Vitamin D is a fat soluble vitamin and resembles with hormones in structure and functions. It is also regarded as a sun-shine vitamin, as it is synthesized in the body from sunlight exposure. The term *vitamin* D_1 was originally used for fat-soluble crystalline form of vitamin D, but now it is used for a mixture of lumisterol and vitamin D_2(1:1).

Chemistry of Vitamin D

They refer to a group of fat-soluble secosterols which occurs as Provitamin D_2 (Ergosterol) in plants and Provitamin D_3 (7–dehydrocholesterol) in animals. The two main forms of vitamin D are Vitamin D_2 (Ergocalciferol) and Vitamin D_3 (Cholecalciferol) (Fig. 3.5), collectively known as calciferol.

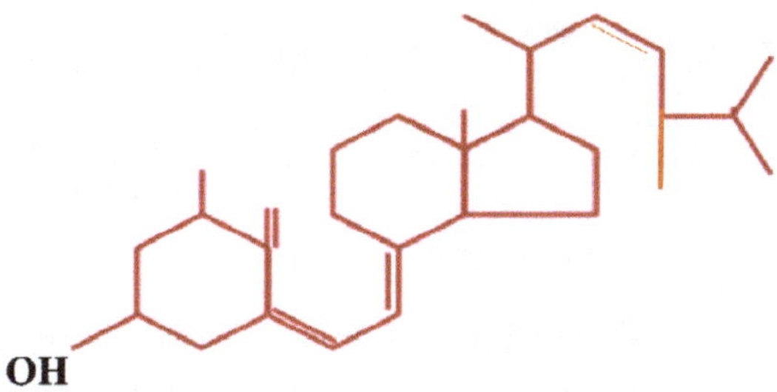

Vitamin D2 (Ergocalciferol)

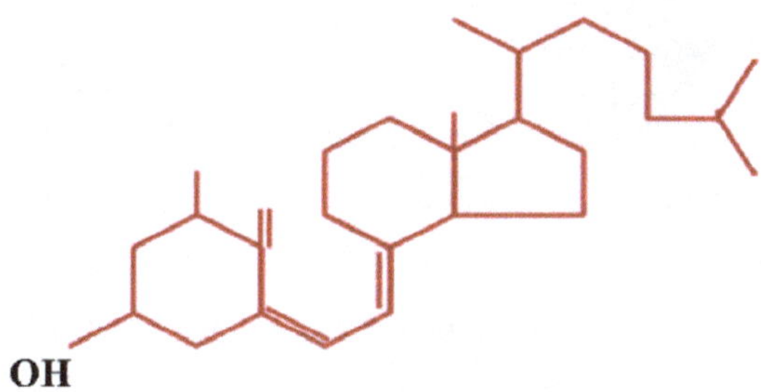

Vitamin D3 (Cholecalciferol)

Fig. 3.5: Physiological forms of vitamin D

Biochemical functions

Vitamin D acquired through food or skin synthesized, is biologically inactive. The active form is made in liver and kidney after enzymatic conversion. UV irradiation of ergosterol that appears in yeast, molds and higher plants synthesize vitamin D_2. Likewise, UV(B) irradiation photosynthesize vitamin D_3 from the 7-dehydrocholesterol, present in vertebrates skin. In liver, vitamin D_3 (cholecalciferol) is changed into calcifediol (25-hydroxycholecalciferol); similarly vitamin D_2 (ergocalciferol) is changed into 25-hydroxyergocalciferol. In the kidney, calcifediol further hydroxylated to make calcitriol, also called as 1,25-dihydroxycholecalciferol (Fig. 3.6). Calcitriol is the biologically active form. It diffuses as a hormone in the blood, having a primary role in controlling the concentration of calcium and phosphate, hence, boost modeling of bone and growth. Vitamin D is not a needed dietary factor as synthesized adequately by most mammals due to enough sunlight exposure. Thus, it is technically considered as a hormone rather than a vitamin.

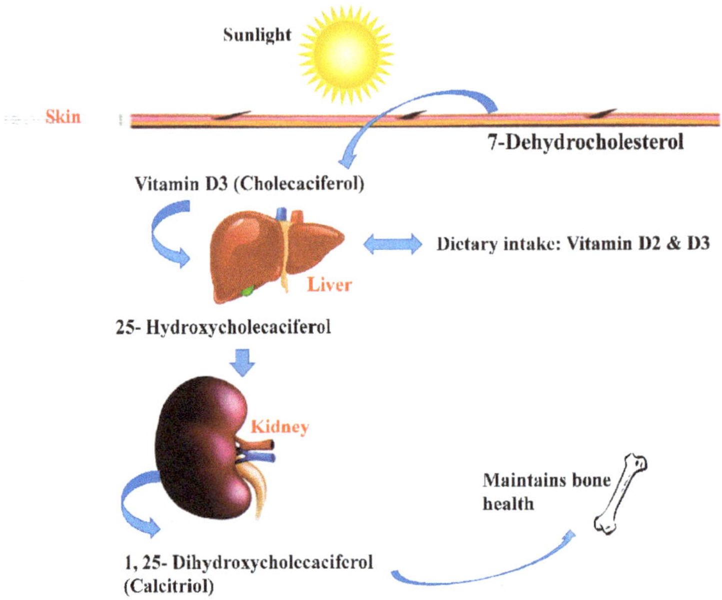

Fig. 3.6: Formation of vitamin D and its role

Dietary sources

The good source of vitamin D includes fatty fish liver oils and egg yolk. Milk is also a source of vitamin D. Cow milk contains 2 IU of vitamin D as per USDA Nutrient database information.

It can be enriched in the body through:

a) Exposure of skin to sunlight for vitamin D synthesis.

b) Consumption of irradiated foods like yeast that contains precursors of vitamin D.

c) Fortification with vitamin D in foods like milk, butter etc.

Deficiency manifestations

Vitamin D deficiency is typified by faulty mineralization or by demineralization of the skeleton. Its deficiency correlates to lack of sun exposure and consumption of food lacking it or due to many chronic illnesses. The skeletal disease includes rickets, osteomalacia, osteoporosis and non-skeletal diseases including autoimmune diseases, inflammatory bowel disease, infections, immune deficiency, cardiovascular diseases and neurocognitive disorders.

Among children, vitamin D lacking is a common cause of bone deformities like rickets which is characterized by faulty mineralization, resulting in soft and pliable bones and lag teeth formation. Therefore, the weight bearing bones are not able to carry body weight and bent to form bow-legs. Vitamin D deficiency in adults results in osteomalacia, which is a demineralization defect in the skeleton that increases susceptibility to fractures, secondary hyperparathyroidism and osteoporosis. In chronic renal failure, it is characterized as renal rickets due to decline in the level of calcitriol synthesis in the kidney.

Hypervitaminosis D: Among fat-soluble vitamins, vitamin D is the most toxic in megadoses (10-100 times RDA). Toxic effects comprise demineralization of bone and more calcium absorption from intestine leading to elevated calcium level in blood plasma. This would on long time results in calcium deposition in tissues like kidney and arteries. This may induce the formation of kidney stones and associated with loss of appetite, nausea, weight loss, increased thirst etc.

Vitamin E

Vitamin E possesses the biological activity of α-tocopherol, found in various oils, whole grains cereals and animal tissue. Its lack in the diet leads to resorption or abortion in female and affects sterility in male animals. Hence, this vitamin named as an anti-sterility vitamin.Vitamin E is synthesized only by plants, thus it is requisite in the diet of animals and human.

Chemistry of Vitamin E

It encompasses faction of tocopherols and tocotrienols, characterized by a 6-hydroxy chromane (tocol) ring, liable for antioxidant property and an isoprenoid side chain (Fig. 3.7). There are about eight isomers of vitamin E, positioning of methyl groups from chroman nucleus designate them as α-, β-, γ-, δ-. Among these, α-tocopherol is the naturally occurring and biologically active form. This is second most common form of vitamin E after γ-tocopherol. Tocopherols and tocotrienols can be distinguished by their phenyl "tails" which is saturated in the tocopherols but unsaturated in the tocotrienols.

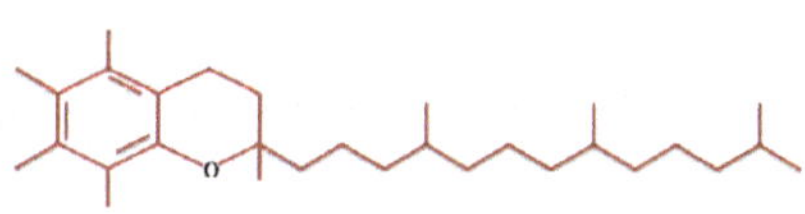

Fig. 3.7: Structure of Vitamin E

Biochemical functions

Major function of vitamin E is to act as non-specific antioxidant. It is soluble in lipid thus accompany lipoproteins, fat deposits, and cell membranes. It prevents polyunsaturated fatty acids (PUFAs), Low-density lipoprotein (LDL), vitamin A, carotenes and other components of cell membrane from oxidation by free radicals. Thus, vitamin E acts as a scavenger and gets itself oxidized into quinone form.

Vitamin E is also needed for proper reproductive functions and prevents sterility by conserving and maintaining germinal epithelium of gonads. It works in association with other vitamins such as vitamin A, vitamin C, and carotene to delay the onset of cataract. This vitamin is also important for membrane structure and integrity. Mostly, vitamin E is found in foods containing fat.

Dietary sources

It occurs primarily in plant products, primarily in vegetable oils and to a little extent, seeds, nuts,and cereal grains. It is also present in milk, butter, eggs, and meat. The tocotrienols appears in rice bran oil, palm oil and the bran and germ fractions of cereals like oat, barley & rice.

Deficiency manifestations

Vitamin E deficiency is rare but it can occur in premature infants. Thus, it is also described as 'vitamin in search of disease'. The deficiency is accompanied by infertility, megaloblastic anemia and degenerative changes in muscle. Among all fat-soluble vitamins, vitamin E is least toxic in overdoses.

Vitamin K

It is a vitamin with coenzyme function and needed for synthesis of prothrombin; blood clotting factors, thus, essential for blood clotting or coagulation (German, Koagulation, hence the name K for the vitamin).

Chemistry of Vitamin K

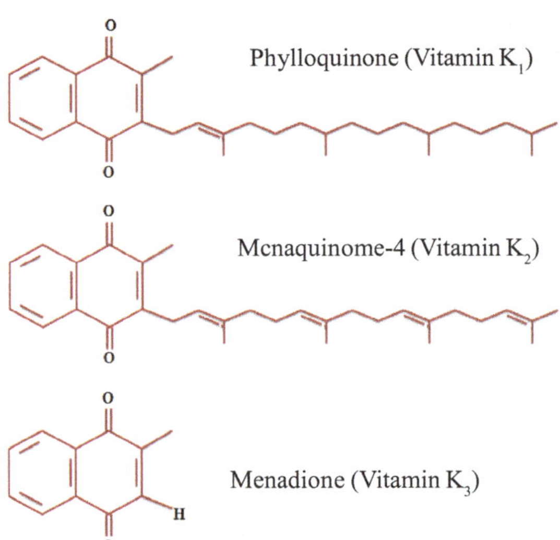

Fig. 3.8: Structure of vitamin K

This is a group of structurally related, lipophilic vitamins needed for complete synthesis of certain proteins that are vital for blood coagulation and involved in metabolic pathways. It is also needed in bone mineralization and cell growth.Chemically, the vitamin K (K_1, K_2 and K_3) is 2-methyl, 1,4-naphthoquinone (Fig. 3.8) and all its derivatives are either taken in diet or synthesized by intestinal bacteria. Vitamin K_1 and K_2 are naturally appearing vitamers. Vitamin K_1 (phylloquinone), the plant form, appears largely in green leafy vegetables because of direct involvement in photosynthesis. It can also be transformed into vitamin K_2 (menaquinone) in animals. Vitamin K_2 further consists of a number of similar chemical subtypes, differing isoprenoid chain. Vitamin K_3 (Menadione) is synthetic form, but no longer used to treat a deficiency as it may be toxic by interfering with the function of glutathione.

Biochemical functions

The prime function of this vitamin is in association with blood clotting process and post-translational modification of some blood clotting factors. The clotting factors II (prothrombin), VII, IX, and X are synthesized in the liver as inactive precursors. Vitamin K acts as a coenzyme for the carboxylation of glutamic acid residues on vitamin K-dependent proteins, catalyzed byenzyme, carboxylase and the transformation of glutamate (Glu) into carboxyglutamate (Gla). These proteins are associated with coagulation, bone and tissue formation and cell growth. Fibrin is a vitamin K dependent protein that clumps together and becomes part of the clot and stops bleeding during injury.

The inhibitors like dicumarol, an anticoagulant present in spoilt sweet clover plus warfarin, which is a synthetic analog, inhibits vitamin K action. They are required to stop blood clots formation preventing hemorrhaging to death, in case of clot lodge in the heart or blood vessels. The protein, osteocalcin modified by vitamin K and enables it to bind to calcium. This calcium is prerequisite for bone matrix.

Dietary sources

Although vitamin K is synthesized adequately in the gut by bacterial synthesis and only half of the body requirement is provided via diet. It is found primarily in green leafy vegetables like spinach, lettuce, cabbage, cauliflower, broccoli, brussels sprouts and absorption increases when associated with a lipid. It also appears in some fruits like grapes, kiwifruit, and avocados. Animal products like milk, egg yolk, fish, meat, liver are the dietary source.

Deficiency manifestations

Vitamin K_2 is synthesized in adults by intestinal bacteria; thus its deficiency is not common among humans. Deficiency may occur due to lack of bile salts or loss in illness. It leads to impaired blood clotting and, potentially, bleeding and blood clotting time increases profusely.

It is relatively toxic in high doses and causes jaundice, brain damage and hemolytic anemia due to increased breakdown of RBC.

Water-soluble vitamins

The water-soluble vitamins include eight well-recognized members of the vitamin B complex: thiamin, riboflavin, niacin, vitamin B_6 (pyridoxine, pyridoxal, and pyridoxamine), biotin, pantothenic acid, folic acid, vitamin B_{12}; and L-ascorbic acid (vitamin C). All vitamin B complex vitamins are essential for enzymatic activity and act in the form of co-enzymes. A list of water-soluble vitamin and their coenzyme forms is shown below (Table 3.1):

Table 3.1: Water-soluble vitamins and their coenzyme forms

Vitamin	Coenzyme form
Viramin B_1 (Thiamine)	Thiamine Pyrophosphate (TPP)
Vitamin B_2 (Riboflavin)	Flavin Mononucleotide (FMN) Flavin Adenine Dinucleotide (FAD)
Vitaman B_3 (Niacin)	Nicotinamide Adenine Dinucleotide (NAD) Nicotinamide Adenine DinucleotidePhosphate (NADP)
Vitamin B_5 (Pantothenic acid)	Coenzyme A (CoA)
Vitamin B_6 (Pyriodoxine)	Pyridoxal phosphate (PALP) Pyridoxamine phosphate (PAMP)
Vitamin B_7 (Biotin)	Biocytin
Vitamin B_9 (Folic acid)	Tetrahydrofolic acid (THFA)
Vitamin B_{12} (Cyanocobalamin)	Deoxyadenosyl cobalamin
Vitamin C (Ascorbic acid)	-

Vitamin B_1 (Thiamine)

Chemistry of Vitamin B_1

Thiamine is a water soluble vitamin initially extracted from an aqueous extract of rice bran. It comprises aminopyridine and thiazole ring coupled with a methylene bridge (Fig. 3.9). It is only naturally appearing compound with thiazole ring. The coenzyme form TPP is formed when the alcohol group of thiamine esterified with two moles of phosphate.

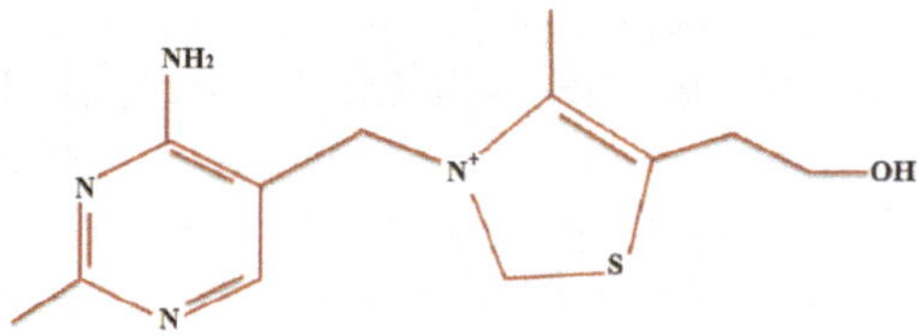

Fig.3.9 Structure of vitamin B_1(Thiamine)

Biochemical functions

Thiamine pyrophosphate (TPP), coenzyme form of thiamin, is metabolically active form. The phosphorylation of thiamine with ATP forms thiamine pyrophosphate (TPP) or diphosphothiamine (DPT). Specific enzymes need TPP as a prerequisite cofactor. TPP is needed in various biochemical reactions engaged in the breakdown of glucose to release energy. In oxidative decarboxylation and transketolase reactions, it acts as a coenzyme, hence an important factor in the metabolism of carbohydrate and branched-chain amino acids in presence of magnesium (Mg^{2+}). In thiamine deficiency, pyruvate or lactate level is elevated in blood. Thiamin also plays a crucial role in nerve functioning.

TPP take part in two principal metabolic reactions:

(a) In α–ketolsformation, catalyzed by transketolase

(b) In α–keto oxidation, by dehydrogenase complexes

Dietary sources

Thiamine appears typically in all plant and animal foods such as cereals, heart, liver, and kidney. The outer layer of cereals (seed) is rich in thiamine. In yeasts and animal tissues, thiamine is present chiefly as thiamine pyrophosphate (TPP). Milk has comparably small amount of thiamine.

Deficiency manifestations

High mortality rate and impairment due to ingestion of polished rice in certain countries caused by the thiamine deficiency, remained a concerned disorder until 1930s in South-East Asia. Beri-beri occurs due to lack of thiamine in diet, acknowledged by dry (paralytic) and wet (oedematous) forms. If lactating mother lack this vitamin, it is likely that newborn would also be deficient for the vitamin. The high carbohydrate diet primarily from milled rice and foods with anti-thiamin factors cause beriberi in adults. The clinical symptoms of beriberi are anorexia, weight loss, apathy, dyspnea, short-term memory, confusion, loss of muscle function, loss of sensation and cardiovascular diseases such as an enlarged heart and increased rate.

Wernicke Korsak off syndrome also known as **cerebral beri-beri** is frequently found in chronic alcoholic individuals with lesser food consumption also. Although thiamin deficiency is rare now, some part of the populations could be on marginally or sub-marginally deficient. Signs are less likely in sub-clinical deficiencies like tiredness, headache and reduced productivity.

The overall decline in the carbohydrate and amino acid metabolism (via α–keto acids) causes reduced synthesis of acetylcholine, a neurotransmitter essential for neurological functions.

In illness like gastrointestinal or liver disease, absorption of thiamine also declines.

Raw seafoods like fishes and mollusks have thiaminase (an enzyme) which breaks down thiamine in the body. Hence, individual ingesting such foods may show symptoms of thiamine deficiency.

Vitamin B_2 (Riboflavin)

The name riboflavin in comes from the word *'ribitol'*, reduced form of ribose sugar and parts in structure and 'flavin' (from Latin *flavus*, 'yellow'), the ring-moiety which gives yellow color to the oxidized molecule. It is also widely known to be yellow enzyme due to its color. The reduced form, which appears in metabolism together with the oxidized form, is colorless.

Chemistry of Vitamin B_2

Riboflavin is water soluble yellow-orange colored substance. It is chemically 7, 8-dimethyl-10 (1'- D-ribityl) isoalloxazine (Fig. 3.10). It is widely distributed all over the tissue in the animal body as flavin but little occurs as free riboflavin. It is heat stable, plus stable against acid and oxidation. However, light particularly ultraviolet light can destroy this vitamin. Thus, it is necessary to protect food (for example, milk) from sunlight having even moderate amounts of riboflavin. As it is heat stable, very little amount is lost during cooking. It acts as a precursor for coenzymes FMN (Flavin mononucleotide) and FAD (Flavin adenine dinucleotide). Flavoproteins are the enzymes that need FMN or FAD as cofactors. Some of these may have metal ions and are called metallo-flavoproteins. Riboflavin functions as a catalyst via these flavoco-enzymes for redox reactions in many metabolic pathways and thereby in energy production. The reduced forms of FMN and FAD participate in enzymatic reactions accompanying flavo proteins, $FMNH_2$ and $FADH_2$, respectively.

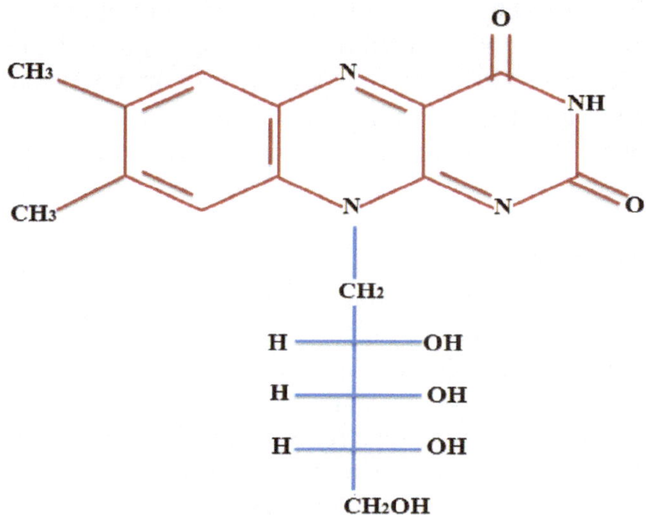

Fig. 3.10 Structure of vitamin B_2 (riboflavin)

Biochemical functions

Riboflavin serves as the precursor for the coenzymes which are essential for energy releases such as dihydrolipoyldehydrogenase and succinate dehydrogenase. It takes part as both one- and two electron transfers in various metabolic reactions or oxidation-reduction reactions and in energy generation via the respiratory pathway.

The two coenzyme forms, FMN and FAD serve as carriers for hydrogen and releases energy via metabolism of carbohydrates and fats as adenosine triphosphate (ATP). It assists each cell in utilizing oxygen more adequately, thus, vital in cell respiration. FMN & FAD helps in pyridine nucleotide reliant or non-reliant dehydrogenation reactions, hydroxylation, deoxygenating reactions with sulfur comprising compounds, oxidative decarboxylation, and reduction of O_2 into hydrogen peroxide with subsequent removal of a hydrogen ion from substrates.

Riboflavin is also peculiar for growth, cell metabolism, tissue respiration and needed for transformation of pyridoxine to pyridoxal phosphate. In presence of an enzyme, flavokinase riboflavin phosphorylates, changed into FMN which is prerequisite for lipid biosynthesis.

Dietary sources

All plants, most bacteria, yeasts, and molds are capable of synthesizing riboflavin. Leafy vegetables, legumes, milk, cheese, eggs, legumes, liver, kidney, and heart are largest repositories of this vitamin. The riboflavin amount in cow's milk is about five times than the human milk. Fruits and most of the root vegetables are the average sources. The milled flour contains low amount, thus the white flour too. However, its content rises steeply in germinated cereals. Ashbyagossypii, a yeast, makes it in such massive amounts that crystals of riboflavin appear in the culture medium. Intestinal microflora also synthesizes riboflavin in gut.

Deficiency manifestations

Deficiency of riboflavin may appear as cheilosis (fissured, cracked and sore lips), eye disorders, mouth and tongue inflammation, weakness, fatigue and skin lesions accompanied with other signs like keratitis, conjunctivitis photophobia and anemia. However, these signs are not specific to ariboflavinos is since alike symptoms may be visible during deficit of nicotinic acid and iron. Moreover, patients with pellagra and beriberi disorders are seen to be deficient in riboflavin. Riboflavin deficiency often appears with other B-complex vitamin deficiencies.

Vitamin B_3 (Niacin)

Vitamin B_3 refers to niacin or nicotinic acid and also termed as pellagra preventive (PP) factor.

Chemistry of Vitamin B_3

The nicotinic acid (pyridine 3-carboxylic acid), nicotinamide (nicotinicacid amide) and the other coenzyme forms of this vitamin altogether referred to as niacin (Fig. 3.11).

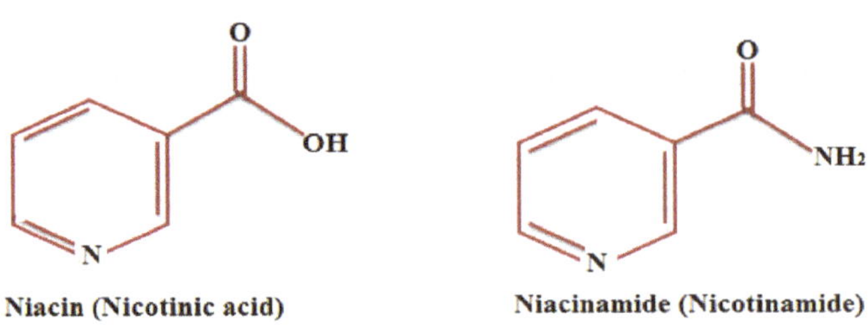

Fig. 3.11: Structure of vitamin B_3

Biochemical functions

In kidney, brain as well as liver, the transformation of nicotinic acid into niacinamide occurs in presence of glutathione. The amidation of nicotinic acid adenine dinucleotide forms the nicotinamide, which subsequently leads to the formation of NAD. In human and other animals, niacin is endowed from the tryptophan that also cures pellagra. However, certain substances like pyridine-

3-sulfonic acid and its amide inhibit growth and it can be recapitulated by supplementing niacin.

The vitamin's coenzyme forms, NAD and NADP, carry out two prime functions in the tissues:

(a) Oxidation of amino acids, aldehydes, alcohols, and hydroxy-carboxylic acids

(b) Reduction of the flavin coenzymes

The active form of niacin is nicotinamide, which serves as a constituent of the coenzymes, NAD (nicotinamide adenine dinucleotide) and NADP (nicotinamide adenine dinucleotide phosphate). NAD acts as an electron carrier in an intracellular respiration, while NADP acts as a hydrogen donor in the reductive biosynthesis of fatty acid and steroid. Both NAD and NADP are a cofactor for enzymes. Being a part of these two coenzymes, niacin takes parts in most of the redox reactions and activates around 200 dehydrogenases, prerequisite in electron transport and other cellular respiratory reactions. In non-redox reactions, NAD is utilized as a substrate by two classes of enzyme that segregate the nicotinic acid moiety from NAD and gives ADP-ribose to proteins. This molecule stimulates calcium ions release from internal repository and thus, plays a crucial role in cell signaling.

Dietary sources

Nicotinic acid appears in plants, animal tissues and yeasts, primarily as nicotinamide. The other sources are meat, milk, and eggs. Cereals have little or no niacin content. However, these foods are good pellagra-preventive sources due to high tryptophan. Fruits and vegetables are not good sources of niacin. Since it is heat stable and not oxidized easily, chances of losses are little during heating, processing and cooking. Like thiamine, most of the niacin is reduced by milling.

Deficiency manifestations

Its' deficiency induces pellagra in man and black tongue in dogs. Pellagra is typified by 3 "Ds", namely dermatitis, diarrhea and dementia. The initial symptoms of pellagra are dubious like thickening of the skin, anorexia, fatigue, inflammation of mouth and tongue, burning sensations and numbness. The severity appears when the disorder is accompanied by the presence of parasites or chronic disorders in children. Clinical signs area loss of appetite, headache, diarrhea, fatigue, emotional instability or amnesia. These may further cause changes in the skin, stomach, intestinal tract, mucosa of the mouth and the nervous system which is characterized as raw skin; hence the name 'pellagra'and are much evident in sun-exposed parts of the skin. Other signs are constipation or diarrhea, gastric mucosa dizziness, vomiting and tongue inflammation. If untreated, pellagra is deadly.

Vitamin B_5 (Pantothenic acid)

The pantothenic acid literally means 'from everywhere',likely named because it is obtained practically from all food sources.

Chemistry of Vitamin B_5

In animals, pantothenic acid is a prerequisite for the synthesis of its coenzyme form i.e., coenzyme A (Co~A or CoA). It is also involved in synthesis and metabolism of several proteins, carbohydrates, and lipids. A pantothenate is an anionic form of this vitamin. Nearly all pantothenic acid, derivative form or its precursors are a supplement in foods and beverages and are chemically synthesized. D-pantothenic acid is water soluble, viscous and yellow colored, heat-labile vitamin and retains biological activity. Calcium pantothenate form of this vitamin is much stable and typically appears in dietary supplements. The disulfide form, pantethine is contemplated to be most active due to the existence of sulfhydryl-group prerequisite for biological activity in CoA. Pantothenic acid comprised of two components, pantoic acid and β-alanine conjoined by a peptide bond.

The coenzyme A (Fig. 3.12) form encompasses pantothenic acid joined to the β-mercaptoethanolamine (thioethanolamine) at one end. At the other end, vitamin is attached to adenylic acid via phosphate bridging. The adenylic acid is made nitrogenous base adenine and ribose-3-phosphate.

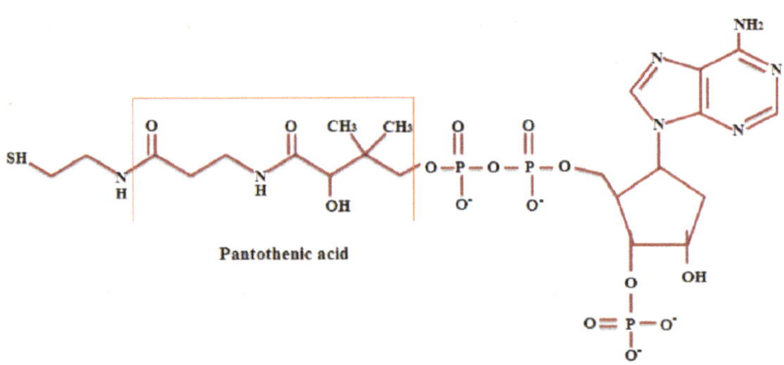

Fig. 3.12 Structure of coenzyme A

Biochemical functions

Pantothenic acid is existent in coenzyme A & acyl carrier proteins (ACP), conveying acetyl and acyl groups, respectively. Coenzyme A is a prerequisite cofactor in lipid elongation, fatty acid synthesis as well as their oxidation. It takes parts in evolution of many secondary metabolites being steroid molecules (for example, steroid hormones, vitamin D and bile acids), poly-isoprenoid containing substances (for example, dolichol, ubiquinone, squalene, and cholesterol), acetylated compounds (for example, acetylated derivatives of amino sugars, acetylated neurotransmitters), prostaglandins and similar compounds. Coenzyme A is needed amid biosynthesis of phospholipids, plasmalogen, sphingenin, and ceramide. It candidly participates in disruption of the carbon skeleton in nearly all amino acids and pyrimidine bases.

ACP is associated with fatty acid synthases, polyketide synthases, lysine synthesis, and non-ribosomal peptide synthetases. Nearly all plants and microorganisms carry out biosynthesis of the vitamin enzymatically by conjoining pantoic acid and ß-alanine. However, the mammals lacking the enzyme for this step are unable to synthesize this vitamin.

Dietary sources

Pantothenic acid can be produced by green plants and several microorganisms but not by mammals. Peanuts and peanut butter, almonds, wheat bran, cheese, liver, kidney, and lobster are good source of this vitamin. The processing of food like refining, freezing, canning, and cooking causes a significant reductionin processed foods.

Deficiency manifestations

Pantothenic deficiency is extremely rare as it is present almost in all foods. However, its deficiency induces several health issues like weakness or fatigue, abdominal pain, respiratory infections, nausea, vomiting, headaches, irritation, insomnia, depression and burning feet syndrome (tingling sensations). The content of vitamin declines in alcoholics, women on oral contraceptives, individual not able to absorb vitamin due to certain diseases.

In rats, pigs, dogs, and chicks, deficiency induces hair or feather de-pigmentation, atrophy of the adrenal cortex with necrosis and hemorrhage.

Vitamin B_6 (Pyridoxine)

The compounds other than thiamine and riboflavin, capable of preventing dermatitis in rats are entitled as Vitamin B_6. Henceforth, they are also called as adermin or antidermatitis factor.

Chemistry of Vitamin B_6

Vitamin B_6comprised of pyridoxine, pyridoxal, and pyridoxamine (Fig. 3.13). Pyridoxine was first evidenced in 1938, from yeast and liver. Later, other two forms were also discovered. All the three forms of vitamin B_6 are derivatives of pyridine and could be distinguished from each other form different groups present at 4th position of the ring involved in pyridine's structure.These three forms are readily inter-convertible into each other biologically. Pyridoxine is a white crystalline compound soluble in water and alcohol and sparingly soluble in lipid solvents. Pyridoxine is sensitive to light and ultraviolet irradiation but heat stable in both acidic and alkaline conditions; in contrast, the two other forms pyridoxal and pyridoxamine are impaired by high temperatures. Coenzyme pyridoxal phosphate (PLP) is the active form of this vitamin and may be formed from all three forms.

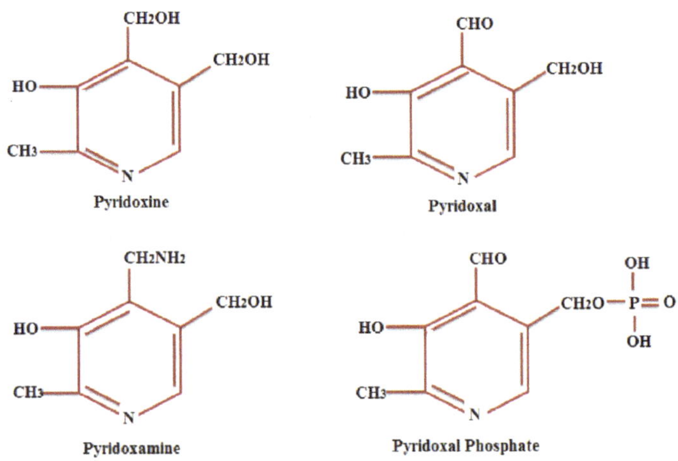

Fig. 3.13: Structure of Vitamin B_6

Biochemical functions

All three forms of vitamin B_6 act as growth factors for numerous bacteria. Pyridoxine can be transformed into pyridoxal or pyridoxamine but none of them can be changed into pyridoxine. Additionally, these forms can change into pyridoxal-5-phosphate that serves as a coenzyme in various enzymatic reactions. These are also occupied in the metabolism of amino acid, fatty acids, glycogen and decarboxylation & racemization reactions. Moreover, metabolism of tryptophan, hydroxyl, and sulfur-containing amino acids also needs vitamin B_6. They probably serve as a carrier for effective movement of amino acids across cell membranes. All these forms are substantially expelled out in urine as 4-pyridoxic acid.

Pyridoxal phosphate (PLP) acts as coenzyme for both glutamic decarboxylase and γ-aminobutyric acid transaminase enzymes. These enzymes are prerequisite for normal metabolism in brain, effective movement of amino acids across cell membranes and formation of arachidonic acid from linoleic acid, Vitamin B_6 deficiency also results in disruption of glycine metabolism.

Dietary sources

Unlike, other B vitamins, this vitamin is widespread in plant and animal tissues. They are exceptionally rich in cereals (wheat, rice), chickpeas, brussels sprouts, pistachios, potatoes, sweet potatoes, bananas, avocados, and yeasts. It also appears in egg yolk, turkey, chicken, fish, beef, pork, and liver. It sufficiently appears in human and cow's milk.

Deficiency manifestations

Deficiency of Vitamin B_6 (also known as apyridoxosis) in rats promotes development of a disease known as acrodynia and characterized by dermatitis on ears, mouth and tail with edema and scaling of these parts. In human infants, a manifestation of this vitamin induces confusion, conjunctivitis, anemia, dermatitis and gastrointestinal problems like nausea and vomiting. Furthermore, tryptophan metabolism is also disturbed. However, deficiency occurs rarely because the intestinal bacteria are capable of synthesizing this vitamin. In B_6-deficient anemia, the RBCs becomes microcytic and hyperchromic. Some malabsorption diseases such as celiac disease may also promote vitamin B_6 deficiency. Dietary deficiency may also be seen in infants, alcoholics, and women taking oral contraceptives.

Vitamin B_7 (Biotin)

This vitamin is also recognized as anti-egg white injury factor or vitamin H. This was extracted as the crystalline substance by a Dutch biochemist named Fritz Kogl. The name biotin came from the name 'bios' given to it by the discoverer. It was then demonstrated that biotin is compatible to the "anti-egg white injury factor" which is liable for treatment of egg white injury brought about in rats and other animals by feeding them raw egg white. The raw egg white comprises of avidin, which is a biotin-binding protein.

Chemistry of Vitamin B_7

Its structure comprises of two rings namely imidazole and thiophene, fused together with a fatty acid side chain as depicted in Fig. 3.14. Allobiotin and epibiotin are its two existing forms. Biotin is an optically active compound. The D&L forms are partly active compared to naturally occurring biotin. The oxybiotin, in which oxygen atom takes the place of S atom, has certain activity. Biocytinis considered as the coenzyme form of the vitamin. It is a bound form joined with the amino acid lysine. It can be formed by intestinal bacteria in such a noticeable amount that the content excreted surpasses the intake.

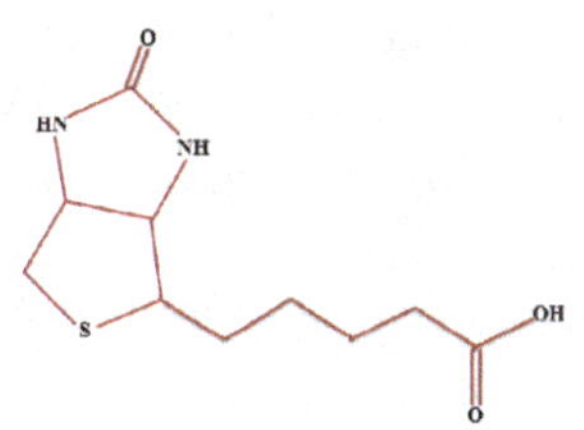

Fig. 3.14: Structure of Vitamin B_7 (Biotin)

Biochemical functions

It makes part of several enzymes by making covalent bond with them in form of prosthetic group. Mainly, the enzymes with biotin as prosthetic group catalyze carboxylation reactions, thus fixing CO_2 into organic substances. The CO_2 is carried as a carboxyl group, joins to one of the ureidonitrogen atoms of biotin, making N-carboxybiotin complex. This vitamin is also involved in the synthesis of fatty acids like oleic acid.

Dietary sources

It appears largely in both plants and animals. Milk, yeast, kidney, liver, and molasses are excellent sources of biotin. Peanuts and eggs have little content of biotin.

Deficiency manifestations

The intestinal bacteria, present in most animals including humans, produce an appreciable amount of biotin in gut. Hence, its deficiency is rare in humans. However, deficiency may be brought about by consuming raw egg white or in diseased person which is characterized by signs like dermatitis, rashes, conjunctivitis, loss in weight, edema, and neurological issues. Avidin (in raw egg white protein) binds with biotin in gut and makes it unavailable to the body.

Vitamin B_9 (Folic acid)

Lucy Wills and co-workers found out that autolyzed yeast preparations were quite effective in curing megaloblastic anemia in pregnant women. Other researchers also performed similar studies on monkeys (with induced anemia) and found yeast extract effective against anemia. They assigned the name 'vitamin M' and 'vitamin Bc' to them. Later on, in 1941, Mitchell and his co-workers isolated and purified a compound from spinach leaves, exhibiting properties of an effective growth factor for rats and *L. casei*, and named it 'Folic acid'. Eventually, in 1946, when the structure of folic acid was confirmed,it was found out that all the substances discussed above were actually folic acid. Moreover, numerous closely related substances or factors have also been extracted from various sources with distinct biochemical functions like *Streptococcus lactis R (SLR) factor*, *fermentation Lactobacillus casei factor*, *Bc conjugate* and *citrivorum factor (CF).*

Chemistry of Vitamin B_9

Different derivatives of pteroic acid exhibiting vitamin B_9 activity are commonly represented by the term 'Folate'. Folate's parent compound 'Folic acid' consists of two rings namely glutamic acid and p-aminobenzoic acid, joined together by a methylene bridge. A glutamic acid side chain is also associated with p-aminobenzoic acid (Fig. 3.15). The glutamic acid present in the structure is also referred to as pteroyl glutamic acid (PGA). Its active form, tetrahydrofolate (THF or FH_4) is produced in response to action of dihydrofolate reductase on folic acid. The folic acid and other substances closely related to it differ from each form in number of glutamic acid groups attached. For example, the *fermentation lactobacillus casei factor* has three glutamic acid residues. *Citrivorum factor* differs from other forms due to the existence of a ring of the pterin moiety. The *Bc conjugates* have seven residues. Certain species does not retain enzyme conjugase, vital for the release of free vitamin. In such species, vitamin B_9 exists in dormant state.

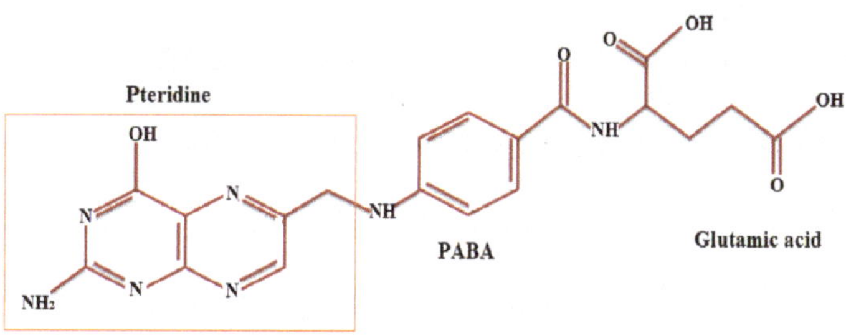

Fig. 3.15: Structure of vitamin B_9

Biochemical functions

Folic acid is first reduced into dihydrofolic acid (DHFA or FH_2) by enzyme folic reductase and subsequently into 5,6,7, 8 tetrahydrofolic acid (THFA or FH_4) by dihydrofolic reductase. These reduced forms of folic acid function as a coenzyme. The formation of FH_4 needs ascorbic acid and is correlated to oxidation of NADPH or NADH.

Folic acids as coenzyme are involved in purine & pyrimidine synthesis and one-carbon metabolism. THFA functions as an acceptor of a one-carbon unit in metabolism in various synthetic reactions. It also moves methyl group and utilizes single carbons in the formation of amino acids (serine, methionine), nitrogenous bases (thymine, adenine, and guanine) and choline. In presence of ascorbic acid, the transformation of folic acid into citrivorum factor takes place (as in tyrosine metabolism) and is known as folicinic acid. This citrivorum factor (CF) supports growth of *Leuconostoc citrivorum*, which is thousand times more biologically active than folic acid. This factor is related with the release of a substance that speedup RBCs production rate. The micro-organism which uses PABA as growth promoting factor also needs folic acid.

Dietary sources

Folic acid is widespread in nature. Green leafy vegetables are an important source of this vitamin. A few other major sources are liver, yeast, kidney, seafood, salmon, tuna fish, asparagus, wheat, dairy, dates and beetroot. Root vegetables, sweet potatoes, bananas, tomatoes, corn, rice, and lamb contains little amount of it. Milk is a poor source of folic acid.

Deficiency manifestations

Folic acid deficiency is considered very common, especially in pregnant women and infants. In vitamin B_9 deficiency, purine synthesis declines due to formation of deoxythiamidine monophosphate (dTMP), which further impairs the DNA synthesis. Due to this, the development of erythrocytes is impaired leading to macrocytic anemia, characterized by macrocytic RBC as a result of impairment in cell division in bone marrow. Its deficiency in pregnant women may cause neural defects in the fetus. Thus, the high doses of this vitamin are recommended by doctors to pregnant women in order to prevent birth deformities. Rats develop achromotrichia (fault in the natural hair color) in vitamin B_9 deficiency. Recent studies also revealed that folic acid is also protective effect against Alzheimer disease.

Vitamin B_{12} (Cyanocobalamin)

Cyanocobalamin is a water soluble vitamin, crystalline in nature and red in color. It is levorotatory and stable against heat in neutral solutions but its activity can be impaired by high temperature and in acidic or alkaline conditions. Hodgkin established the complex structure of vitamin B_{12} through X-ray crystallography in 1955. Pernicious anemia, characterized by bright red smooth tongue and tingling & numbness in hands and feet is a characteristic feature of vitamin B_{12} deficiency. Therefore, this vitamin is also called as anti-pernicious anemia factor or factor X.

Chemistry of Vitamin B_{12}

The structure of this vitamin is remarkably complex one (Fig. 3.16). The exceptional feature of this vitamin is the existence of cobalt in the trivalent state into its molecule. No other cobalt-comprising organic substance exists in nature. The cobalt atom is located centrally and is enclosed in a macro cyclic structure of four reduced pyrrole rings unanimously termed as corrin. It is evident from the structure that the six valences of the cobalt atom (CO^{2+}) are contented by the four nitrogen of the reduced tetrapyrrole, a nitrogen atom of 5, 6-dimethylbenzimidazole and a cyanide ion. The two pyrrole rings are coupled directly thus, the corrin has a lesser degree of unsaturation with only 6 double bonds. Then the other two pyrrole rings, are attached with single methene carbon. Another apparent feature of the structure is the existence of a loop of the isopropanol, phosphate, ribose, and 5,6-dimethyl-benzimidazole in such a way that end of the loop is being linked to the cobalt atom. It is the remarkably common form and is occasionally penned as vitamin B_{12}a. In other forms of this vitamin cyanide ion is interchanged with other ions, For example, in hydroxycobalamin (vitamin B_{12}b) with hydroxyl ion, in nitrocobalamin (vitamin B_{12}c) with nitrite ion etc. The B_{12}b and B_{12}c can be transformed into vitamin B_{12}a by using cyanide. The coenzyme form of Vitamin B_{12} is analogous to cobalamin apart from the existence of adenosine in place of cyanide ion. Its coenzyme form is only known bio-molecule with a carbon-metal bond.

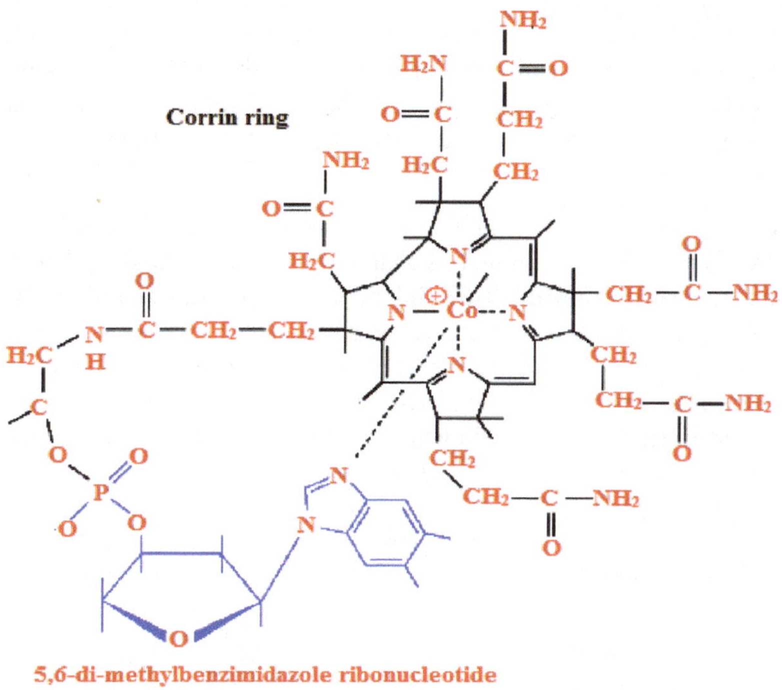

Fig. 3.16: Structure of vitamin B_{12}

Biochemical functions

Vitamin B_{12} exists in two metabolically active forms namely, methyl cobalamine and deoxyadenosylcobalamin or cobamide coenzyme. In methylcobalamine form, methyl group is directly attached with cobalt while in deoxyadenosylcobalamin, adenosine is attached with cobalt. Vitamin B_{12} can be interchanged to its coenzyme forms by extracts from microorganisms enriched with ATP. Deoxyadenosylcobalamin is involved in intramolecular rearrangement of substrates through transfer of hydrogen and carbon atoms. For example, rearrangement of glutamic acid is carried out to form L-threo-β-aspartic acid. Similarly, ethylene glycol is converted into acetaldehyde. Deoxyadenosylcobalamin also catalyzes inter-conversion of methyl malonyl CoA and succinyl CoA. Reduction of ribonucleotides to deoxyribonucleotides is an essential step for DNA synthesis. This reduction takes place through deoxyadenosylcobalamin. Similarly, methylcobalamin coenzyme form is involved in methionine synthesis (from homocycteine), methane formation (from methanol) and acetate synthesis (from CO_2). It also takes part in protein synthesis, reviving-SH enzymes and induction of myelin formation.

Dietary sources

Vitamin B_{12} is obtained exclusively from animal sources such as liver, milk, eggs, meat, fish, clams, and oysters. It may be synthesized by the intestinal microorganisms under certain circumstances. It is absent in natural plant foods except spirulina, a blue-green algae. Animals and plants are not capable of synthesizing this vitamin. It is unique in the context that can be synthesized only by anaerobic bacteria.

Deficiency manifestations: The well-known disorder in connection to its deficiency is pernicious anemia. It is characterized by diminishing hemoglobin level, a decrease in erythrocytes and neurological manifestations. Dietary deficiency is mostly seen in strict vegetarians. The most cases of deficiency are a result of inefficient absorption of the vitamin. The inefficient absorption may be due to absence of an intrinsic factor that should be present in gastrointestinal track for its binding and absorption. In severe cases, irritability, anorexia, and listlessness also occur. The patient weakens, losses weight and nervous system is also affected by its deficiency. Further, severe deficiency can result in demyelinization of the mega nerve fibers of the spinal cord, which ultimately causes death.

Vitamin C (Ascorbic acid)

Vitamin C is a water soluble vitamin, first separated from paprika plant in crystalline form and named as hexuronic acid. Later on, Szent-Györgyi and Haworth identified its antiscorbutic properties and proposed its new name as 'L-Ascorbic acid', which was officially accepted in 1965. As its structure is very much similar to hexose sugars, therefore, this vitamin can also be synthesized from glucose in laboratory. The synthetic form is equally biologically active as the natural one (L-form). It is involved in the treatment of scurvy, therefore, also termed as an antiscorbutic factor.

Chemistry of vitamin C

Structurally, vitamin C is represented by two forms namely, ascorbic acid and dehydroascorbic acid. The latter form arises from oxidation of ascorbic acid. Dehydroascorbic acid can reduce back to ascorbic acid form (Fig. 3.17). Both the forms are biologically active. The chemical name for vitamin C is 1-threo-2,4,5, 6-pentoxyhexen-2- carboxylic acid lactone. The enolic group present at third carbon atom is responsible for acidic properties of vitamin C. It is a colorless, odorless and crystalline compound with little sour taste. L-isomer of the vitamin has antiscorbutic properties. It gets oxidized notably in presence of copper, iron and oxygen. Hence, food cooked in copper appliances loses ascorbic acid rapidly. It is expeditiously damaged by alkalies but is somewhat stable in weak acidic condition. Therefore, baking soda has a detrimental effect on ascorbic acid. Drying of vegetables and their storage culminate a loss of vitamin C.

Biochemical functions

All higher plants and mammals except humans, other primates, and guinea pig can synthesize ascorbic acid from L-gulonolactone. For example, the rat can synthesize this vitamin by themselves and therefore, will never be deficient in vitamin C. In animals, the liver and adrenals (cortical portion) are the prime sites for its synthesis.

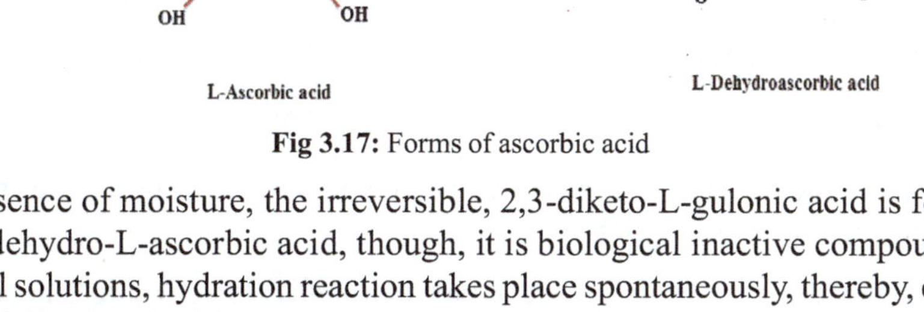

Fig 3.17: Forms of ascorbic acid

In presence of moisture, the irreversible, 2,3-diketo-L-gulonic acid is formed from dehydro-L-ascorbic acid, though, it is biological inactive compound. In neutral solutions, hydration reaction takes place spontaneously, thereby, causes its oxidation i.e., biological inactivation of the vitamin.

A larger function of ascorbic acid is structuring tissue collagen (a protein). Hydroxyproline exclusively appears in collagen, and is a prerequisite for maintaining its tertiary structure. Vitamin C is vital for the activity of the enzyme, collagen proline hydroxylase, which converts proline to hydroxyproline. It also has a role in the transformation of folic acid into a tetrahydrofolic acid (a physiologically active form). It also acts as a specific coenzyme in biological oxidation system. It enhances body's immune system by the releasing interferon, prostaglandins, T-lymphocytes, and immunoglobulins. However, mega doses of vitamin C can leach out calcium and other needed minerals and acts as a diuretic and laxative.

Dietary Sources

The acerola fruit (Malpighiapunctifolia) provide approximately 1,000–4,000mg of ascorbic acid per 100g and is the richest known source of this vitamin. However, it is found in all fresh fruits and vegetables, particularly citrus fruits such as orange, lemon, lime, gooseberry, pineapple, guavas, tomatoes, etc. It is also present in new potatoes, melons, raw cabbage. Green pepper is also rich in vitamin C. Dry seeds and cereals are usually deficient in vitamin C, but sprouting or germination increases its content. Human milk contains approximately 3 mg/100ml vitamin C, which is slightly higher than that of cow milk (2.37 mg/100 ml).

Deficiency manifestations

The primary deficiency manifestation of the vitamin is scurvy with clinical symptoms such as:

a) Tender bones especially noticeable in legs. Legs presume the 'frog like position' appropriately called pseudo-paralysis, often confused with paralysis.

b) Edematous swellings: swelling palpates at the end of the femur.

c) Petechial hemorrhages: The condition upturns red and purple spots (or petechiae) on the skin and mucous membrane, owing to the bursting of capillaries.

d) Bleeding gums: Typified by bluish-purple spots, swellings of the mucous membrane and bleeding in gums.

e) Scorbutic rosary: The sharp and angular beaded structure, formed at costochondral junctions

f) Delayed wound healing: The wound healing is slowed or does not appear because of the failure of the cells to make collagen fibrils.

g) Cessation in bone growth: The cells of growing epiphyses continue to escalate but, the new matrix does not develop. As a result, bones fracture easily at the time of growth or may totally cease down.

h) Sicca syndrome: Swelling in joints and follicular hyperkeratosis may occur, which is normally correlated with collagen disorders and comprises keratoconjunctivitis sicca.

i) Anemia: low iron content and impairment of folic acid metabolism

j) Pyrexia: Infantile scurvy

Vitamin-Like Substances

Besides the vitamins explained earlier, there are many other substances present in food as accessory factors. They may not be prerequisite in diet but perform important functions and are regarded as vitamin-like substances such as choline, inositol, lipoic acid, PABA, carnitine, and bioflavonoids.

Choline

Choline is a water-soluble, vitamin-like vital nutrient. It is a component of phospholipid (lecithin) which is a part of several plants and animal organs. It is trimethylhydroxy ethanolamine and can be made by the body. It is existent in dietary sources like milk, egg, liver, cereals etc. Choline avoids the aggregation of fat in the liver by the disposal of triacylglycerol from liver and promotes synthesis of phospholipids and lipoproteins. It also serves as a precursor for the formation of acetylcholine, a neurotransmitter which is essential for transmission of nerve impulse.

Inositol

It is also named as myo-inositol or meso-inositol, which is basically hexahydroxy-cyclohexane. This is a prerequisite for the synthesis of phosphatidyl-inositol (lipositol) which is part of many cell membranes. It acts as a lipotropic factor, thus prevents fat accumulation in the liver along with choline. It also acts as a secondary messenger for certain hormones and accompanies the release of Ca^{2+} ions at the membrane level. However, its hexaphosphate, phytin found in many plants and prevents absorption of iron and calcium from the intestine.

Lipoic acid

It is also known as thiotic acid basically 6,8-dithioctanoic acid, a sulfur-containing fatty acid. It exists in both oxidized and reduced form. It is soluble in fat as well as water. It is involved in decarboxylation reactions along with vitamins like thiamine, niacin, riboflavin and pantothenic acid. It is regarded as a universal antioxidant. It is found to reduce the free radicals in the brain that otherwise contributes to Alzheimer's disease and multiple sclerosis. Lipoic acid triggers the production of glutathione (GSH).

Para-aminobenzoic acid (PABA)

4-aminobenzoic acid (PABA) is a part of folic acid as a constituent. It may be referred as a vitamin in another vitamin (folic acid). The lack of PABA induces failure in lactation and graying of black hair in rats. It is synthesized by bacteria and is a prerequisite for their proliferation. The antimicrobial sulfa drug, sulfanilamide and related compounds compete with PABA for making a place in coenzyme i.e., vital for bacterial reproduction thus, constrain their proliferation.

Carnitine

Carnitine, derived from an amino acid, is found in nearly all organisms and animal tissue. Carnitine's role is correlated with the deportation of fatty acid from the bloodstream within muscle cells to active sites of fatty acid oxidation, thereby administered the rate of fatty acid oxidation. It is synthesized nearly in all organisms; therefore, it is not regarded as a true vitamin.

Bioflavonoids

They are flavonoids, initially isolated from lemon and thought to prevent scurvy and responsible for maintenance of normal capillary permeability. The term vitamin P (P for permeability) was used for this group of substances. There are around 4000 flavonoid compounds that are characterized and classified, but all are not approved by FDA.

Antivitamins

Antivitamins are antagonistic to the action of vitamins. They usually have structural similarities with vitamins. Various vitamins and their antivitamins are shown below in Table 3.2.

Table. 3.2: List of antivitamin or antagonist along with the corresponding vitamin

Vitamins	Antagonist or antivitamin
Vitamin K	Dicumarol Warfarin
Vitamin B_1 (Thiamine)	Thiaminase Pyrithiamine Oxythiamine
Vitamin B_2 (Riboflavin)	Galactoflavin
Vitamin B_3 (Niacin)	Isoniazid Deoxypyridoxine
Vitamin B_7 (Biotin)	Avidin Desthiobiotin
Vitamin B_9 (Folic acid)	Aminopterin Methotrexate Trimethoprim

Sample questions

1. Define vitamin. Give detailed classification of vitamins.
2. Enlist fat-soluble vitamins. Discuss their toxicity in brief.
3. Which enzyme converts the carotene into retinol and how?
4. What is the role of vitamin A in visual cycle?
5. Discuss the sources, functions and dietary deficiency disease of vitamin A.
6. What do you understand by vitamers and its role in the body?
7. Why is vitamin D considered as a hormone?
8. What are the major functions of vitamin D in the body and mechanism of action for formation of an active form of vitamin D?
9. Discuss the sources, functions and dietary deficiency disease of vitamin D.
10. Give a brief account of vitamin E.
11. What are the different forms of vitamin K and its role in blood clotting?
12. Enlist different types of water-soluble vitamins and their coenzyme forms, if any.
13. Give a brief account of the active form of thiamine and explain the coenzyme role of it.
14. What are the deficiency manifestations of vitamin B_1?
15. Briefly explain the structure of vitamin B_{12}.
16. What is egg white injury? Which vitamin is linked with it?
17. Write a short note on vitamin C forms.
18. Discuss in brief deficiency manifestations of vitamin C.
19. What are vitamin-like substances? Give a brief account of different vitamin-like substances.

Suggested Readings

Bender, D.A. (1992). *Nutritional Biochemistry of Vitamins*. New York: Cambridge University Press.

Berg, J. M., Tymoczko, J. L., Gatto, G. J. and Stryer, L. (2015). *Biochemistry*. 8thedn. New York: W.H. Freeman & Company

Bolander, F.F. (2006). 'Vitamins not just for enzymes',*CurrOpinInvestig Drugs*. 7, pp. 912.

Combs, G. F. (2008). *The Vitamins Fundamental Aspects in Nutrition and Health*. 3rdedn. San Diego, CA: Elsevier Academic Press.

Cox, M. M., and Nelson, D. L. (2008). *Lehninger Principles of Biochemistry*. 5thedn. New York: Worth Publishers.

Eskin, N. M. A. and Shahidi, F. (2013). *Biochemistry of foods*. 3rdedn. USA: Academic Press Elsevier.

Gaby, S. K., Bendich, A., Singh, V.N. and Machlin, L. J. (1991). *Vitamin Intake and Health*. New York: Marcel Dekker.

Joint, F.A.O. and World Health Organization (1998). *Vitamin and mineral requirements in human nutrition*. 2ndedn. Bangkok: WHO Library.

Linder, M. C. (1985). *Nutritional biochemistry and metabolisms with clinical applications*. New York: Elsevier Science publishing company Inc.

Powar, C. B. and Chatwal, G. R. (2006). *Biochemistry*. 5thedn. Mumbai: Himalaya Publishing house.

Rucker, R.B., Suttie, J.W., McCormick, D. B. and Machlin, L.J. (2001). *Handbook of vitamins*. 4thedn. New York: Marcel & Dekker Inc.

Sardesai, V. (2012). *Introduction to clinical nutrition*. 3rd ed. USA: CRC Press (Taylor & Francis Group), pp. 157-164.

Satyanarayana, U. and Chakrapani, U. (2013). *Biochemistry*. 4th ed. Kolkata: Elsevier Books and Allied (P) Ltd.

Singh, N.P. (2005). *Biochemistry*. New Delhi: Akansha Publishing House.

Srilakshmi, B. (2009). *HumanNutrition (for B.Sc. Nursing students)*. New Delhi: New age International (P) Ltd., Publishers.

Swaminathan, M. (1985). *Advanced textbook on Food nutrition*. Volume I Fundamental aspects. Bangalore: The Bangalore press

Talwar, G. P., Srivastava, L. M. and Moudgil, K. D. (1989). *Textbook of biochemistry and human biology*, 2ndedn. New Delhi: Prentice Hall of India Pvt. Ltd.

Voet, D., Voet, J. G. and Pratt, C. W. (2016). *Fundamentals of biochemistry: life at the molecular level*. 5thedn. Hoboken, NJ: Wiley.

Zempleni, J., Suttie, J. W., Gregory III J. F. and Stover, P. J. (2014). *Handbook of Vitamins*. 5thedn. Parkway, WA: CRC Press (Taylor & Francis Group).

4

Hormones

Introduction

Hormones may be defined as specific chemical molecules produced in certain specific organs called specialized cells or endocrine glands. They are delivered into the blood stream or some other body fluid in minute amounts by the circulatory system of the body and their they exerts a physiological effect at a site of action remote from its origin, regulating numerous biological activities including growth, development, reproduction, water & electrolyte balance, storage & usage of energy etc.

Difference between hormones and vitamins

1. Hormones are mostly secreted by endocrine ductless glands and some by neuro-secretary cells of animals while vitamins are synthesized by plants or by microorganisms and then taken by animals mostly through diet.
2. Chemically, hormones are either water soluble amino acids or polypeptides or fat soluble steroids in nature while vitamins are various compounds like esters, organic acids, amide, amine, alcohol etc.
3. Hormones never acts as coenzymes, they may be excitatory or inhibitory in nature while vitamins are mostly coenzymes or constituents of coenzymes.
4. Hormones directly influence gene expression while vitamins have catalytic action.
5. Lesser than the required secretion of hormones causes metabolic disorders while deficiency of vitamins causes specific deficiency diseases.

Difference between hormones and enzymes

1. Hormones are not always proteins in nature while enzymes are mostly protein (except ribozymes).
2. Hormones promote or inhibit processes like growth, flowering, sex expression etc. while enzymes are biological catalysts and control all biochemical reactions of the cells.
3. Hormones are mobile in nature, means they are produced at one site and show their activities at other site while enzymes are present in all cells and show their activities their only.
4. Hormones get used up after their participation in biological processes while enzymes remains unchanged at the end of the reaction and may be reutilized.

Properties of hormones

1. They are chemical messengers, secreted by some specialized cells called endocrine cells to regulate the behavior of target cells.
2. Their secretion is regulated by nervous system through feedback effect.
3. They are low molecular weight compounds. Therefore, they can pass through blood capillaries easily.
4. They are produced in very low concentration.
5. They are antigenic in nature.
6. Whenever their function is over, they are readily destroyed or inactivated.

Classification of hormones

The main endocrine glands secreting hormones are located in three different regions in humans as shown in Fig. 4.1.

a. Hypothalamus and pituitary in the head region

b. Thymus, thyroid and parathyroid in neck region

c. Adrenal, pancreas, ovaries and testis in the abdominal region.

However, the hormones from different endocrine glands can be classified on the traditional basis, structural basis and on the basis of mode of action as shown in the Fig. 4.2.

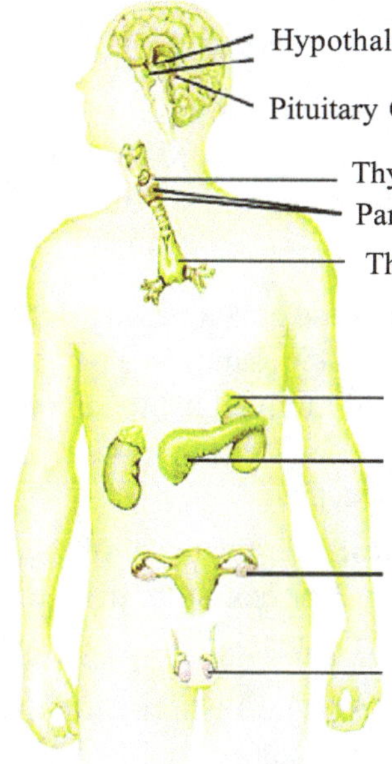

Fig. 4.1: Location of different endocrine glands in humans

(1) Traditional classification

(a) ***Classical hormones:*** These hormones are secreted by the endocrine glands and are distributed in all body parts by circulatory system.

(b) ***Neuro hormones:*** These hormones are secreted by neuro endocrine cells at the nerve terminals and transported by blood vessels.

(c) ***Local hormones:*** These hormones are secreted by the cells into the interstitial fluid and are of further two type-Paracrine hormones and Autocrine hormones. The hormones which acts on the neighboring cells are called as paracrine hormones and those which act on the cell from which they were secreted are called as autocrine hormones.

(2) Structural hormones

(a) ***Steroid hormones:***These hormones are derivatives of cholesterol and are lipid soluble. For example, sex hormones (androgens, estrogens) and hormones secreted by adrenal glands (mineralocorticoids, glucocorticoids). Water balance, sexual development and stress response are the key functions of these hormones.

(b) ***Amino acid derivative hormones:*** As the name indicates, these hormones are derivatives of amino acids like tryptophan and tyrosine. For, example, thyroid hormone is an amino acid derivative hormone which regulates development of organs. Catecholamine (epinephrine and nor-epinephrine) are another neuro and stress hormone which comes under this category.

(c) ***Peptide hormones:*** These hormones are derived from peptides and are stored into the cell vesicles until there is a stimulus for their secretion into blood stream. For example, thyroid stimulating hormone, insulin hormone and prolactin.

(d) ***Fatty acid derivatives:*** These hormones are derived from arachidonic fatty acid and are also called as eicosanoids. They are produced by each cell of the body for regulating inflammation, blood pressure and blood clotting.

(3) On the basis of mode of action

(a) ***Quick acting hormones:*** The response of these hormones is immediate on target organ or cells. For example, protein and amino hormones are usually quick acting hormones.

(b) ***Short acting hormones:*** The response of these hormones is delayed comparatively. These are usually small sized and bind with the protein receptors present in the cytoplasm. For example, adrenal cortex hormones and steroid hormones of reproductive system are short acting hormones.

Glands secreting different hormones have been grouped according to location into following heads:

(1) ***Cranial endocrine glands:*** They lie in head cavity. For example, Pineal and Pituitary

(2) ***Pharyngeal endocrine glands:*** These lie in neighborhood of pharynx. For example, Thyroid and Parathyroid

(3) ***Abdominal glands:*** These lie in abdomen. For example, Pancreas, intestinal glands and adrenals

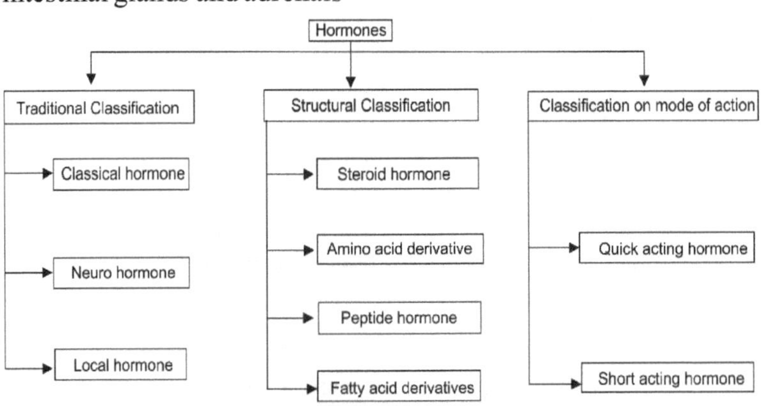

Fig. 4.2: Classification of hormones

Some important hormones

Hypothalamic and pituitary hormones

Hypothalamus is an almond sized structure of diencephalon of the forebrain, situated below the thalamus and just above the brain stem. It acts as a bridge between nervous system and the endocrine system (Fig. 4.3). Its primary functions are to maintain the homeostasis of the body and give directions to the pituitary gland for releasing different hormones as per the needs of the system. The pituitary gland or hypophysis is a bean shaped gland situated in a hanging position through a stem like structure called infundibulum or pituitary stalk. It is further made up of two different parts called posterior lobe of pituitary (neurohypophysis) and anterior lobe of pituitary (adenohypophysis).

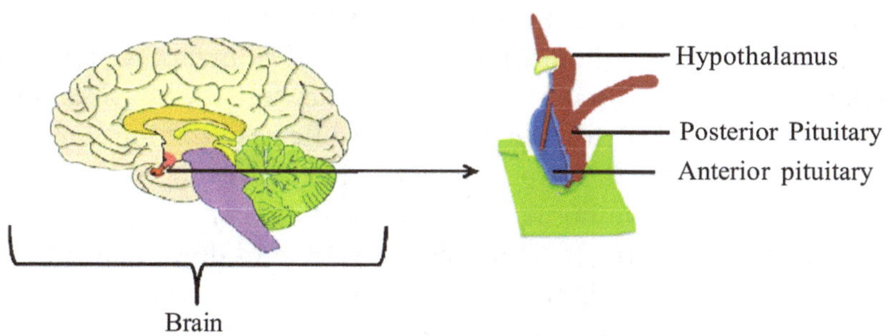

Fig. 4.3: Location of hypothalamus and pituitary glands in brain

Functions:Both hypothalamus and pituitary together act as "Command center" for the endocrine system. Whenever stimulus is received by the peripheral nervous system, the same is allowed to funnel through the hypothalamus-pituitary complex for translation into a hormone. This hormone then either directly initiates the required response or directs the other glands to release the required hormones to initiate the response. The hypothalamus interacts with the two different lobes of pituitary gland in two different ways.

Anterior pituitary Hormones:The interaction of hypothalamus with anterior lobe of pituitary gland is through a network of blood vessels and capillaries called as hypophyseal portal system. The hypothalamus secrets Gonadotrophin Releasing Hormones (GnRH) into the anterior lobe of the pituitary through hypophyseal portal system, where it stimulates the release of Follicle Stimulating Hormone (FSH) and Luteinizing Hormone (LH). These two hormones then travel down to the respective gonads (testis in males and ovaries in females) and stimulate them to release their hormones. Another hormone released from the hypothalamus into the anterior lobe of pituitary is Corticotropin Releasing Hormone (CRH). CRH stimulates release of Adenocorticotropin hormone

(ACRH). ACRH goes down to the adrenal glands and stimulates them to release their hormones. Similarly, Thyroid Releasing Hormone (TRH) and Growth Hormone Releasing Hormone (GHRH) from hypothalamus stimulates pituitary (anterior lobe) to release Thyroid Stimulating Hormone (TSH) and Growth Hormone (GH), respectively. TSH goes down to stimulate the thyroid gland to release further thyroxine and tri-iodothyronine. GH goes down to the long bones & the big muscles and stimulates growth. There is one hormone called as prolactin inhibitory factor which is constantly secreted from hypothalamus into the anterior lobe of pituitary. When it stops being released, the pituitary gland is signaled to release prolactin. Prolactin is involved in milk production in lactating women.

Posterior pituitary Hormones:Hypothalamus interacts with the posterior lobe of the pituitary through stimulation nerves which run down the pituitary stalk and causes release of a number of hormones. The main difference between anterior and posterior pituitary hormones is that anterior lobe is stimulated by hypothalamus to secrete its own hormones while posterior lobe only stores the hormones released by the hypothalamus and releases them in response of a stimulus. The two main hormones of the posterior pituitary are Antidiuretic Hormone (ADH) and oxytocin hormone. ADH goes down and stimulates the collecting ducts of the kidneys for maximizing the water retention. Oxytocin is involved in uterine contractions in women.

Thyroid hormone

Thyroid hormones are secreted by a 2 inches long, butterfly shaped gland situated at the base of the neck below the thyroid cartilage and in front of the trachea (wind pipe), called as thyroid gland. The gland consists of two lobes left and right connected by a thin bridge of thyroid tissue, called as isthmus. Thyroid releases mainly two hormones T_3 (triodothyronine) and T_4 (thyroxine), normally in the ratio of 20:80. Thyroid also produces calcitonin but to a lesser extent. Calcitonin helps in controlling blood calcium level. More than 50% of the total body iodine is present in thyroid gland as it is essential component of the structure of thyroid hormones. Both T_3& T_4 are formed by iodination of tyrosine amino acid as shown in the Fig.4.4.The thyroid gland captures iodine from blood capillaries again through concentration and electrical gradient using ATP dependent transport mechanism and concentrates it in 30-50 times more amount than in the circulation.

Synthesis of thyroid hormones: Both T_3& T_4 are formed by iodination of tyrosine amino acid as shown in the Fig.4.4. Hydrogen peroxide oxidizes iodine to the active form using thyroid peroxidase enzyme. This active iodine gets incorporated into a large glycoprotein called Thyroglobulin. This iodinated thyroglobulin undergoes a series of coupling reactions. Two tyrosine residues, each iodinated at two positions couple together to synthesize thyroxine (T_4) whilst one di-iodotyrosine combines with one mono-iodotyrosine to synthesize tri-iodothyronine (T_3). Thyroglobulin when hydrolyzed yields thyroxine, common amino acids and iodinated derivatives. Thyroxin was first isolated by kendall in 1919.

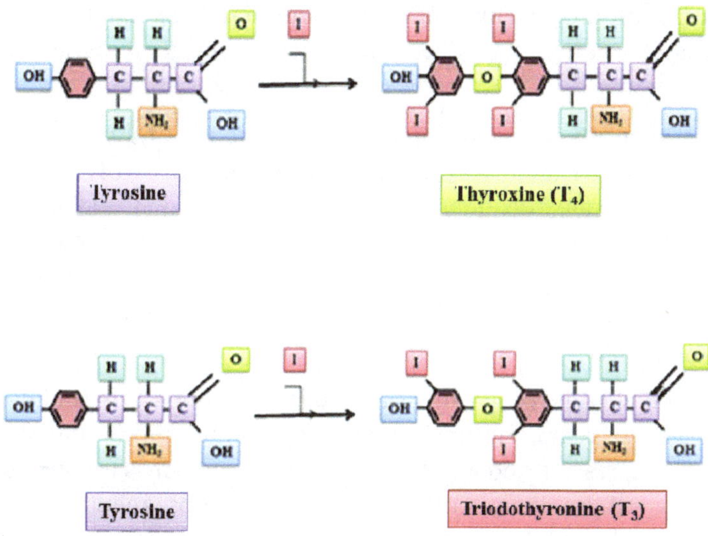

Fig. 4.4: Synthesis of thyroid hormones

Functions

1. Thyroid hormones increase the rate of the metabolism in different tissues by means of increasing the size and number of mitochondria (cell powerhouse), stimulating the synthesis of enzymes involved in respiratory chain and increasing the permeability of cell membrane for sodium and potassium ions.
2. They stimulate transport of sugar, calcium ions and ATPase activity.
3. They increase protein catabolism in skeletal muscles.
4. They increase the erythropoietin synthesis in kidneys. Erythropoietin is a glycoprotein required for red blood cell production.
5. They help in increasing the motility and uptake of nutrients from gastrointestinal track.
6. They are also essential for development of nervous system and maturation of bones.
7. In liver, they stimulate gluconeogenesis and glycogenolysis.

Deficiency disorders: Disorders associated with thyroid can occur at any stage due to disease, injury or deficiency of iodine in diet. The main disorders include goiters, hyperthyroidism and hypothyroidism.

Goiters: Intense enlargement of the thyroid gland indicates the goiter disorder. This enlargement of gland starts putting pressure on the wind pipe (trachea) and food pipe (esophagus) present in neck region. The patient feels difficulty in breathing and swallowing. Therefore, these enlargements are also often removed.

Hyperthyroidism: It is a condition of over production of thyroid hormones by over reactive thyroid gland. As the thyroid hormones are involved in controlling metabolism, heart rate, breathing, body temperature, weight and nervous system so the patient feels nervousness, weight loss, anxiety, irregular rapid heartbeats, frequent sweating, mood swings, skin dryness and sleep problems. The autoimmune disorder called as 'Grave's Disease' is one of the prominent causes of hyperthyroidism.

Hypothyroidism: It is a condition of lower production of thyroid hormone by thyroid gland. Its symptoms are reduction in cold toleration power, tiredness, constipation, depression and weight gain. Hypothyroidism is usually not detected normally. Hypothyroidism is often associated with pregnancy and can cause delayed growth and intellectual development in child.

Besides this, solitary thyroid nodules, thyroid cancer and thyroiditis are also some of the common thyroid disorders.

Insulin Hormone

Insulin was first isolated from dog's pancreas by Banting and Best in 1922. In 1926, Able and co-workers succeeded in crystallization of insulin and also established it a protein in nature. Insulin was the first hormone found to be a protein. The chemical structure of insulin was determined by Sanger and co-workers after a hard work of painstaking decade (1944-1954), for which he was awarded with Nobel Prize in 1958.

Structure: The molecular formula of insulin is $C_{254}H_{377}O_{75}N_{65}S_6$. There are two polypeptide chains in the structure of bovine insulin, chain A & chain B (Fig. 4.5). Chain A is made up of 21 amino acid residues and chain B of 30 amino acid residues. Both the chains are connected at two separate points by two disulfide bonds through cysteine residues. Cysteine residue at 7th position of chain A is linked with cysteine residues at 7th position of chain B through disulfides linkages. Similarly, cysteine at 20th position of chain A is linked with cysteine at 19th position of chain B.The hormone losses its biological activity if these disulfide bonds are reduced by action of alkali or by some other method. The molecular weight of the hormone is 5733KD. The isoelectric pH of insulin is 5.4. Zinc is also found in the vicinity of this hormone. However, it is not a part of the hormone's structure. There is a slight variation in the structure of insulin from different species. Sanger and Smith determined the amino acid sequence of insulin form five different species (cattle, sheep, pig horse & whale) and found that there was difference only in 8th, 9th, and 10th position in chain A.

Insulin is biosynthesized in β-cells of pancreas from Proinsulin by lysosomal proteolytic enzymes through cleaving a peptide chain of 33 amino acids from it. When someone is insulin deficient then this hormone is administered intravenously by injection as it gets destroyed by digestive enzymes.

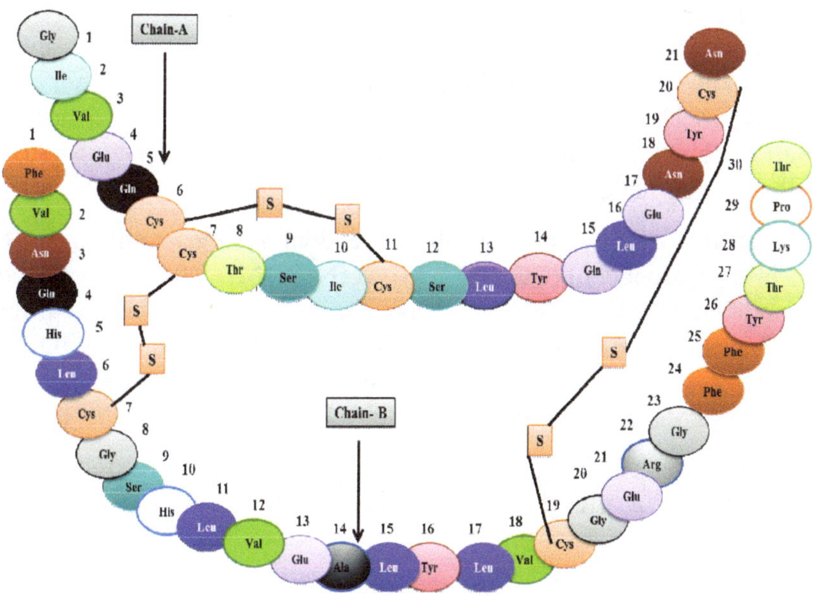

Fig. 4.5: Structure of insulin hormone

Biochemical Functions: Carbohydrate metabolism is intensely affected by insulin.

1. It promotes the entry of glucose and other sugars in to the cells by enhancing penetration of cell membranes and boosting glucose phosphorylation. As a result of this, the blood sugar level falls. Due to this function, insulin is also called as hypoglycemic factor.

2. It assists in incorporation of amino acid residues to form protein (proteogenesis). At the same time, it also acts as anti-proteolytic factor by decreasing breakdown of protein in tissues.

3. Inul in administration also promotes lipogenic effect.
4. Inorganic metabolism is also affected by insulin. It lowers the blood phosphate levels and promotes phosphate (inorganic) uptake by cells. This phosphate is used by the cells as ATP.
5. It stimulates the activity of enzyme hexokinase and glycogen synthase.
6. It stimulates oxidative phosphorylation of mitochondria present in muscle cells.

In general, it can be concluded that the insulin promotes anabolic processes like synthesis of glycogen, fatty acids & proteins and demotes catabolic processes like fat and glycogen breakdown.

Deficiency disorder: The insufficient synthesis or excessive destruction of insulin may lead to a medical complication known as Diabetes Mellitus, characterized by increase in blood sugar, appearance of sugar in urine, ketone body formation in blood, frequent urination (polyuria) especially in night (nocturia), dehydration (polydipsia), unrelenting thrust, excessive eating (polyphagia) and feeling less energetic due to loss of body proteins.

Diabetes mellitus is of two type-insulin dependent diabetes mellitus (IDDM) and non-insulin dependent diabetes mellitus (NIDDM). IDDM occurs due to destruction of β-cells somehow and therefore, the insulin produced is not sufficient enough. This is also called as Brittle Diabetes or Juvenile Diabetes. NIDDM is caused by the fact that the cells become resistant to native insulin. This disorder is also called as Stable Diabetes or Adult-onset Diabetes.

As the disease proceeds, the person becomes weak and losses weight, leading to reduction in plasma volume and kidney failure, ultimately resulting in death of the person.

Adrenal gland hormones

There are two small, triangular or semilunar shaped adrenal glands in human body. They are loosely attached to the kidneys, beneath the peritoneum, enclosed in the connective tissue and protected by a fatty tissue. Morphologically, gland is partitioned in two different components. The central part is called as adrenal medulla and the peripheral part is called as adrenal cortex (Fig. 4.6). Medulla represents 20% of the gland and cortex 80%. The cortex is further divided into three distinct zones, each secreting different hormones to perform specific physiological functions. The outermost layer of the adrenal cortex is called as Zonaglomerulosa. The middle and inner layer are termed as Zonafaciculata and Zonareticularis, respectively. Both parts; adrenal cortex and adrenal medulla, secrete hormones which are chemically and physiologically different from each other. Therefore, the hormones from these two regions of adrenal gland are discussed here separately.

Adrenal medulla hormones

Catecholamine hormones, chiefly epinephrine (adrenaline) and nor-epinephrine (nor-adrenaline) are produced by the chromaffin cells present in the core of the adrenal gland (adrenal medulla) and released at the sympathetic nerve endings. The epinephrine and nor epinephrine are non-steroidal in nature and released by adrenal medulla in the ratio of 80: 20. Tyrosine is the precursor of these hormones. Both nephrine and epinephrine are produced in response to physical, chemical, thermal and emotional stress situations. Therefore, they are also called as hormones produced in response to fight, fright and flight (3F).

Biochemical functions

(1) Effect on carbohydrate metabolism: Catecholamine promotes glycogenolysis and gluconeogenesis by increasing glucose release from liver and decreasing its utilization in muscles. Adrenaline promotes glucagon secretion and inhibits insulin secretion.

(2) Effect on lipid metabolism: Both adrenaline and nor adrenaline promote triacylglycerol breakdown in adipose tissue. The free fatty acids produced by this action are released into the blood from where they are utilized by heart and muscles as a source of fuel.

(3) Effect on physiological functions: Cardiac output, blood pressure and oxygen intake are increased by catecholamine. They also cause relaxation of smooth muscles of skeletal, bronchi and gastrointestinal track, dilation of pupil of eye and constriction of some blood vessels.

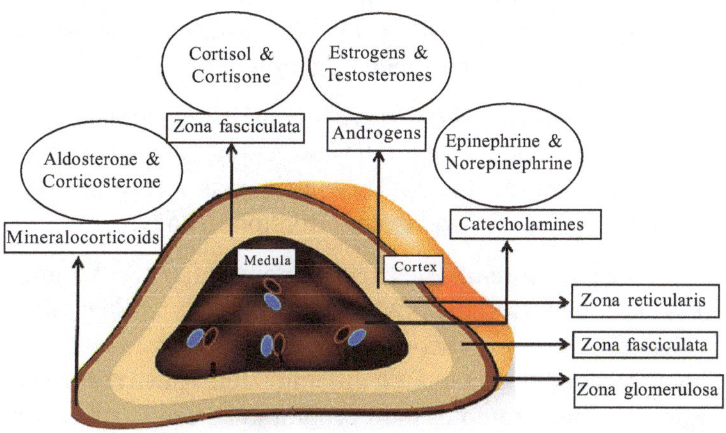

Fig. 4.6: Adrenal gland hormones

Adrenal cortex hormones

The hormones produced by adrenal cortex are steroidal in nature and also classified as C_{21} steroid hormones as they are made up of compounds with 21 carbons atoms in their structure. Adrenal cortex secretes total 40-50 compounds showing hormonal activity and they are also closely related to each-other. Collectively, these hormones are called as corticosteroids. They are actually synthesized from cholesterol in smooth endoplasmic reticulum and mitochondria of endocrine cells. There is no provision for storage of these hormones in vesicles or any other structures of endocrine cells. Therefore, they are diffused into the blood stream as soon as they are produced. These hormones are fat soluble and their receptors are located in the cytoplasm of target cells. Aldosterone (mineralocorticoid), Cortisol (Glucocorticoid) and Dehydroepiandrosterone (DHEA) are the major hormones secreted by zonaglomerulosa, zonafasciculata and zonareticularis, respectively.

Biochemical Functions of Aldosterone: The main biochemical function of mineralocorticoid is in distal convoluted tubules of kidney. They promote reabsorption of sodium ions and send them back into the blood stream. This retention of sodium ions is accompanied by concurrent loss or excretion of potassium ions, hydrogen ions and ammonium ions.

Biochemical functions of cortisol

(1) Effect on glucose metabolism: They promote gluconeogenesis and maintain the blood glucose level balance by coordination with insulin hormone.

(2) Effect on lipid metabolism: They promote lipolysis in adipose tissue and thereby reduce the triacylglycerol levels.

(3) Effect on protein and nucleic acid metabolism: They promote synthesis and transcription of proteins in liver. However, in extra hepatic tissues the action is reverse and catabolic,thereby enhancing protein degradation.

(4) Effect on water and electrolyte metabolism: They regulate the production of arginine vasopressin hormone (ADH). ADH regulates the water balance in blood. In cortisol disorder, the ADH level increases resulting in decreased filtration of water by kidneys and water retention in body.

(5) Effect on immune system: They act as a powerful anti-inflammatory agent and suppress the immune response by affecting the cells participating in immune reactions. Therefore, strong corticosteroids are also used by medical practitioners in case of medical condition involving inflammation. It also reduces lymphocyte multiplication rate and promotes their programmed cell death so that the body remains protected from their overreaction.

(6) Effect on cardiovascular system: They increase the strength of heart muscles for contraction and helps in maintaining sodium potassium balance in heart. They are also helpful in controlling the blood pressure by contraction of arteries.

(7) Effect on central nervous system: Excess or low cortisol levels are responsible for alteration of mood, change in behavior, excitability and electrical activity of neurons. Moodiness, low tolerance power, lack of thoughts clarity and low memory may be due to disproportional secretion of cortisol.

Sex Hormones

The sex hormones are associated with the sexual progressions and the development of differential male and female secondary sex characteristics. In mammals, the sex hormones are synthesized by gonads (ovary in females and testis in males), corpus luteam and the placenta. The activity of these hormones is controlled by the hormones secreted by the anterior lobe of the pituitary. Therefore, the hormones secreted by anterior lobe of pituitary for sending message to the different gonads to secrete sex hormones are also called as primary sex hormones and the sex hormones secreted by the gonads as per the received message are called as secondary sex hormones. Based on number of carbon atoms present, the sex hormones are also classified as C_{18}, C_{19} & C_{21}. Three types of secondary sex hormones are secreted as shown in the Fig. 4.7.

(A) Estrogens (Female or ovarian or follicular hormones)

(B) Androgens (Male or testicular hormones)

(C) Gestogens (Corpus luteal hormones)

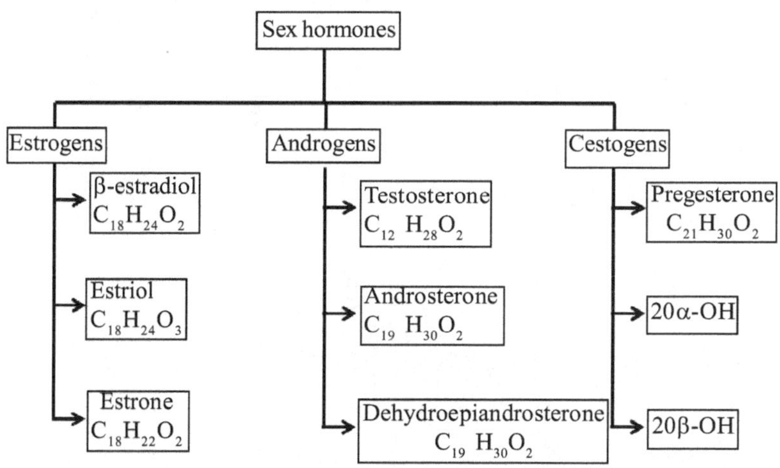

Fig. 4.7: Classification of sex hormones

(A) Estrogens: Ovarian follicles and corpus luteam are the two prominent regions in mammalian ovary to secrete hormones. Estrogens are secreted by follicle region and transported by binding with a specific plasma protein. These hormones are mainly involved in inducing and maintaining the periodic reproductive cycle in females, called as estrus cycle. Chemically, all the estrogens are 18 carbon compounds derived form a hydrocarbon called estrane. There are total three compounds showing hormonal activity, which comes under the group estrogens as shown in the Fig.4.8.

(a) β-estradiol or Dihydrotheelin with molecular formula $C_{18}H_{24}O_2$.

(b) Estriol or Theelol with molecular formula $C_{18}H_{24}O_3$.

(c) Estrone or Theelin with molecular formula $C_{18}H_{22}O_2$.

Adolf Butenandt, a German biochemist and Doisy in 1929 successfully isolated estrone from the urine of pregnant women. Thereafter, estriol was isolated by Marrian followed by isolation of estradiol. Later in 1939, Adoft Butenandt was awarded with Nobel Prize in Chemistry for his discoveries related with sex hormones. Aromatic nature of first ring and absence of methyl group at 10th carbon atom is a characteristic feature of all estrogen hormones. Physiologically, β-estradiol is the most potent followed by estrone and estriol (50: 5: 1, relative potencies). The estrogen levels are highest in young women and then they declines with age. Their declination indicates rise in menopausal symptoms like mood fluctuations, brittle hair, dry skin and decrease in libido. In humans, fat cells also secrete estrone hormones along with ovarian follicles. So the fatty women have an advantage over slim ones.

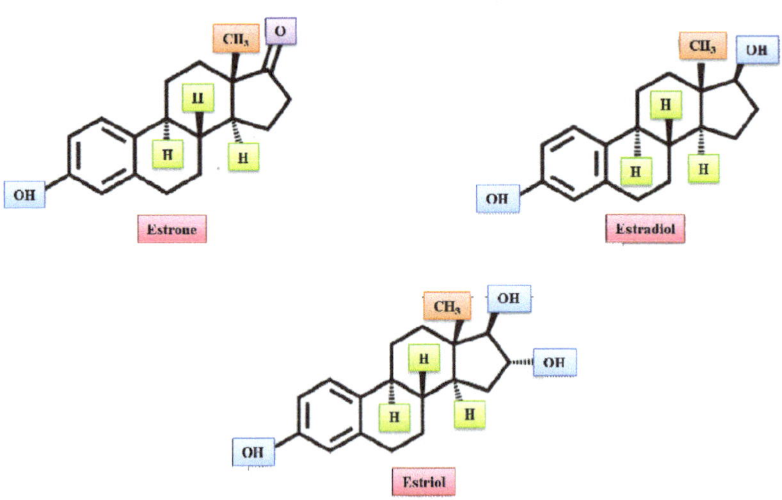

Fig. 4.8: Structure of estrogens

Biochemical Functions

1. Estrogens are responsible for development of secondary sex characteristics like enlargement of uterus and vagina, pelvis widening, periodic vaginal bleeding, proper body and scalp hair ratio, passive and less aggressive attitude, narrow shoulders, broad hips and high pitch voice in females.
2. They help in preparing the uterine mucosa for the future child by inducing proliferation of endometrium, deepening of uterine glands, increasing vascularity and changes in epithelium of vagina & fallopian tubules.
3. They help in maintaining the elasticity of skin and improving memory in post-menopausal women.
4. They slow down the rate of leaching of calcium from bones to blood and hence, protect women from osteoporosis.
5. They help in prevention of cardiovascular diseases by maintaining the flexibility and elasticity of blood vessels.
6. They are used as youth tonic and as one of the main component in hormone replacement therapy along with progestin hormone, for treatment of menopausal symptoms in women.
7. They positively affect hippocampus and cerebral cortex, resulting in enhanced production of choline acetyltransferase, responsible for enhancing brain power.
8. Estrogens also act as antioxidants by scavenging free radicals and also protect brain cells from toxic compounds.
9. They also have been found to have prominent effect on patients with Alzheimer's disease.
10. They also act as cofactor in the trans-hydrogenation between the two pyridine nucleotides NADP & NAD.

(B) Androgens: These sex hormones are mainly secreted by the testis (male sex gonads). Chemically, they are derived from a C_{19} hydrocarbon called as androstane. There are total three compounds showing hormonal activity, which comes under the group antrogens as shown in the Fig. 4.9.

(a) Testosterone with molecular formula $C_{19}H_{28}O_2$

(b) Androsterone with molecular formula $C_{19}H_{30}O_2$

(c) Dehydroepiandrosterone (DHEA) with molecular formula $C_{19}H_{25}O_2$

Testosterone is the principal sex hormone in males. It is synthesized from cholesterol in leydig cells of testis. Cholesterol is converted into pregnenolone, then to progesterone and hydroxyprogesterone through a series of enzyme catalyzed reactions as shown in the Fig. 4.10. The latter is then converted into androstenedione, which is the precursor of testosterone. Four to twelve milligram of testosterone is secreted every day in a healthy male.

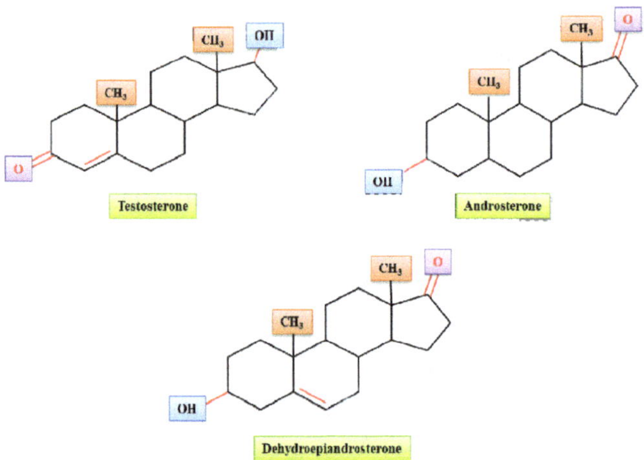

Fig. 4.9: Structure of androgens

Adolf Butenandt and co-workers were the first to successfully isolate androsterone from male urine in 1931. Among all these hormones, testosterone is the most potent followed by androsterone and DHEA (20: 7: 1, relative potencies). Besides these hormones, androst-4-ene-3, 17 dione and androst-4-ene-3, 11, 17 trione are the two testicular hormones secreted by adrenal cortex.

Biochemical Functions

1. Androgens are responsible for development of secondary sex characteristics like increase in the length and width of penis, pigmentation of scrotum, enlargement of seminal vesicles, enlargement of larynx & vocal cord, deepness in voice, anterolateral recede of scalp hair, more aggressive & active attitude, broadening of shoulders with no alteration in hips and thickening of sebaceous gland secretions in males.
2. Testosterone plays its role in muscle building and development. Therefore, it is used for treatment of persons with poorly developed muscles and considered as a youth hormone for males.
3. Testosterone is capable of increasing the size and strength of bones. Therefore, it is used for treatment of osteoporosis, especially in old age.

4. Androgens regulate the activities of male reproductive system.
5. They stimulate the growth of skeletal muscles.
6. They also play their role in male baldness.
7. They inhibit the secretion of anterior pituitary gonadotrophins.
8. They are useful in treatment of metastatic carcinoma of breast.
9. They are also useful in treatment of menopausal syndrome.

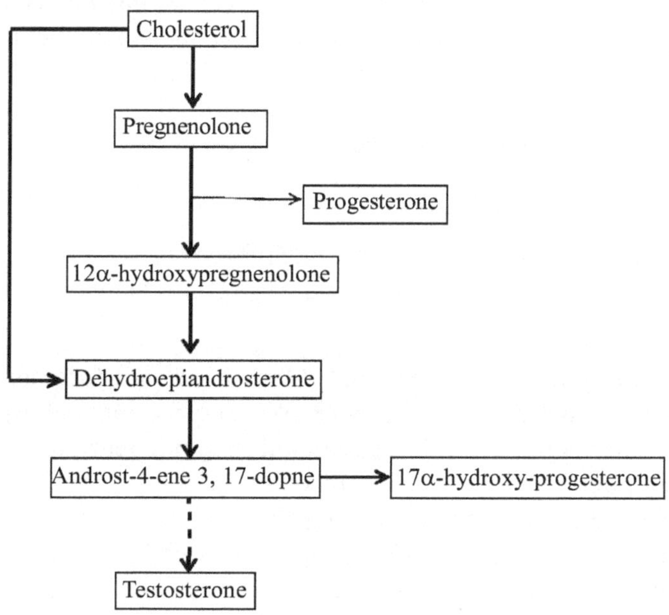

Fig. 4.10: Testosterone synthesis

(C) Gestogens: Gestogens or Progestins are the hormones secreted by the corpus luteam from ovaries. Progesterone is the principal gestogen. Besides this, 20α-OH & 20β-OH are also secreted by corpus luteum. In 1934, Adolf Butenandt and co-workers isolated progesterone in pure form from corpus luteum. Progesterone is biosynthesized from its immediate precursor pregnenolone with the help of dehydrogenase and isomerase enzymes.

Biochemical functions

1. Progesterone promotes the proliferation of uterine mucosa for receiving and implementing the fertilized ova.
2. It develops mammary glands to maturity for breast feeding the infant during lactation period.

3. It also inhibits uterus contractions during pregnancy.
4. It shows an anti-ovulatory effect if given between 5th and 25th day of menstrual cycle. Therefore, it is also used as one of the main component of oral contraceptive.
5. Progesterone is also a precursor of cortisol and corticosterone in adrenal glands.

In 1926, Frederick Hisaw discovered another hormone called relaxin or uterine releasing factor (URF). This hormone is produced during pregnancy. Relaxin is a polypeptide hormone of about 9000 KD molecular weight but its structure is not determined yet as it has not been obtained in pure form. Relaxation of cervix of the pregnant women and inhibition of uterine motility are the two important functions of this hormone.

Mechanism of hormone action

The hormones control the activities of different target tissues by means of specific cellular mechanisms. The five general mechanisms proposed are as mentioned below.

(1) Action at cyclic nucleotide level: Many hormones form an intracellular hormonal mediator called as cyclic 3'-5' adenosine monophosphate (cyclic AMP). The hormones travel through blood stream to reach to the target cells. There they interact with a specific receptor in the plasma membrane, which leads to activation of enzyme adenylcyclase present in the same membrane. The portion of this enzyme protruding towards cell cytoplasm immediately converts cytoplasmic ATP to cyclic AMP in presence of Mg^{2+}. The cyclic AMP then initiates a number of cellular functions like activating the enzymes, altering the cell permeability, intracellular proteins synthesis, contraction or relaxation of the muscles and releasing other hormones (third messengers). After the function is over, cyclic AMP is inactivated by an enzyme called phosphodiesterase. Here the cyclic AMP is considered as second messenger and the main feature of this model is that the hormones do not need to enter inside the target cells and mark their impact on cell membrane only. Cyclic AMP acts as allosteric effector and activates the protein kinases. These activated kinases modify the activities of different cell proteins by phosphorylating them. Other type of intracellular hormone mediators includes cyclic guanosine monophosphate and prostaglandins. These compounds normally cause intracellular inhibition, dissimilar to the activation usually caused by cyclic AMP.

(2) Action at cyclic nucleotide level: In this mechanism, mainly thyroid and steroidal hormones causes synthesis of certain proteins in target cells. These proteins are enzymes, which in turn activate other cell functions. These hormones act at the level of cell nucleus. They bind to specific receptor protein in the cytosol and forms steroid receptor complex. This complex enters into the nucleus and binds to specific sites on the chromatin and influence mRNA synthesis and through that protein synthesis. For example, mineralocorticoids enter into the cytoplasm of renal tubule cells, where they combine with their high affinity specific receptor proteins. This complex then gets the permission to enter in the cell nucleus. After entry, the proteins are synthesized that promote absorption of sodium from renal tubules and potassium secretion into the tubules.

(3) Stimulation of enzyme synthesis at ribosomal level: Some hormones function at the level of information translation carried by mRNA on the ribosomes for production of enzyme proteins.

(4) Hormonal action at the membrane level: There are many hormones in the body which are involved in transport of carbohydrates, amino acids and nucleotides across cell membranes. First they bind with the cell membranes and then cause prompt metabolic modifications in the tissues. Many protein hormones along with adrenaline and nor adrenaline perform their tasks in the same manner. They bind with specific receptors on the cell membrane, stimulating different enzyme systems.

Sample questions

1. Define hormones.
2. Differentiate between hormones and vitamins.
3. Differentiate between hormones and enzymes.
4. Briefly describe properties of hormones.
5. Briefly explain classification of hormones.
6. Discuss hypothalamic hormones in detail.
7. Draw the structure of insulin hormone. Briefly describe its biochemical functions.
8. Briefly explain biochemical functions of adrenal medulla hormones.
9. What are sex hormones? Discuss structure and biochemical functions of female sex hormones.
10. Discuss about the structure and biochemical functions of male sex hormones.
11. Write a short note on various mechanisms of hormonal action.

Suggested Readings

Baniahmad, A. (Ed.). (2002). Thyroid Hormone Receptors: Methods and Protocols (Vol. 202). Springer Science & Business Media.

Birnbaumer, L. and O'Malley, B. W. (1985). Hormone action. Part I. Peptide hormones.

Cox, M. M. and Nelson, D. L. (2008). Lehninger Principles of Biochemistry 5th edn. New York: Worth Publishers.

Eskin, N. M. A. and Shahidi, F. (2013). Biochemistry of Foods. 3rd edn. USA: Academic Press Elsevier.

Linder, M. C. (1985). Nutritional Biochemistry and Metabolisms with Clinical Applications. New York: Elsevier Science publishing company Inc.

Litwack, G. (2015). Hormones and Transport Systems (Vol. 98). Academic Press.

Litwack, G. (2017). Human Biochemistry. Academic Press

Norman, A.W. and Litwack, G. (1997). Hormones. Academic Press.

Satyanarayana, U. and Chakrapani, U. (2013). Biochemistry. 4th ed. Kolkata: Elsevier Books and Allied (P) Ltd.

Schafer, E.A.(1913). A Proposed Classification of Hormones. Proc. 17th Internat. Cong. Med., London", Sec. II, ii, 21, p.10.

Singh, N.P. (2005). Biochemistry. New Delhi: Akansha Publishing House.

Talwar, G. P., Srivastava, L. M. and Moudgil, K. D. (1989). Textbook of Biochemistry and Human Biology, 2nd edn. New Delhi: Prentice Hall of India Pvt. Ltd.

Van der Molen, H. J., King, R. J. B. and Cooke, B. A. (1988). Hormones and Their Actions. Part II, Specific Actions of Protein Hormones (New comprehensive biochemistry; v. 18B). Elsevier Science Limited.

Voet, D., Voet, J. G. and Pratt, C. W. (2016). Fundamentals of Biochemistry: life at the molecular level. 5th edn. Hoboken, NJ: Wiley.

http://biology.tutorvista.com/biomolecules/hormones.html

http://www.playbuzz.com/samchung10/how-well-do-you-know-the-human-anatomy

https://www.slideshare.net/shaguftaakmal/adrenal-hormones-35650035

http://www.cuchd.in/elibrary/resource_library/University%20Institutes%20of%20Sciences/Fundamentals%20of%20Biochemistry/Chap-31.pdf

5

Carbohydrate Metabolism

Introduction

Carbohydrates are the most common source of energy for all living cells, which, in presence of light, are formed by the green plants from CO_2 and water during photosynthesis. It can be said that 'light' is the ultimate source of energy required for all living cells. Glucose, a monosaccharide, plays a central role in carbohydrate metabolism in all plants, animals and many microorganisms. Glucose is derived from the polysaccharides (starch, glycogen) and non-carbohydrate sources (amino acids, glycerols, etc.) during metabolism and ultimately converted either into carbon dioxide and water or other intermediate products. Complete oxidation of glucose to CO_2 and water provides ~2,840 kJ/mol energy. The other monosaccharides such as fructose, galactose, mannose, etc. are also important molecules in carbohydrate metabolism.

Glucose is not only an excellent source of energy; it also acts as a precursor molecule for other biosynthetic reactions such as formation of amino acids, fatty acids, nucleotides, co-enzymes, etc. in some organisms. In cells of the higher organisms (animals and plants), glucose has mainly four fates:

1. Oxidation via glycolysis and citric acid cycle to produce energy.
2. Oxidation via pentose phosphate pathway for nucleic acid biosynthesis.
3. Synthesis of larger polysaccharides like glycogen and starch for storage.
4. Utilization for the synthesis of extracellular matrix and cell wall polysaccharides.

Starting from the food having all macro- and micronutrients, starch and/or other oligosaccharides arc converted into simple sugar molecules, more commonly glucose in small intestine during digestion.Thereafter, glucose is transported into intestinal mucosa through carrier mediated system and moves further towards different cells through blood as a carrier. With the help of transporter proteins or carrier molecules, glucose enters into the cells and above mentioned biological processes (fates 1-4) initiates as per the requirements of the body at that time.

Entry of glucose from blood into the cells

Glucose concentration is very low in human cells as compared to blood. Human blood contains 4.0-6.0 mmol/L (72-108 mg/dL) glucose (on fasting), which can be increased up to 7.8 mmol/L (140 mg/dL) after 2 hours of eating. Nevertheless, glucose does not enter into the cells via diffusion or concentration gradient. It enters into the cells via carrier proteins, which can be divided into two types: 1) Insulin-dependent carrier proteins: for muscles and adipose tissue, 2) Insulin-independent carrier proteins: for hepatocytes (liver), erythrocytes and brain cells. It means hepatocytes, RBCs and brain cells do not require insulin for the transfer of glucose molecules from blood to cells even in diabetic patients.

Glycolysis

Glycolysis is the universal pathway of glucose catabolism in all living cells. Glycolysis (Greek *glykys*-sweet, and *lysis*-splitting), can be defined as a process in which glucose is converted into pyruvate (or lactate) by a series of enzyme-catalyzed reactions and produces energy in the form of ATP and NADH. Glycolysis is also known as Embden-Meyerh of pathway (E.M. pathway) in honor of two biochemists Gustav Embden and Otto Meyerh of, who majorly contributed in the elucidation of this process. The complete pathway was elucidated in 1940.

Gustav Embden
1874-1933

Otto Myerhof
1884-1951

Important features of glycolysis

1. Glycolysis takes place in all living cells of all of the organisms.
2. The site of glycolysis is cytoplasm, where cytosomal enzymes participate in the reactions.
3. Glycolysis can be aerobic or anaerobic depending upon the availability of the oxygen; consequently, resulting in pyruvate or lactate as the end products, respectively.
4. Under aerobic conditions, pyruvate is formed as a result of glycolysis; which, further, is converted in to CO_2 and H_2O after mitochondrial oxidation.

Reactions of Glycolysis

Glycolysis can be divided into two distinct phases:

1. Energy investment phase or preparatory phase
2. Energy generation phase or payoff phase

Energy investment phase or preparatory phase

Preparatory phase involves the breakdown of 6-C glucose molecule into two 3-C glyceraldehyde-3-phosphate molecules by investing 2 ATP molecules. The steps are as follows and shown in Fig. 5.1.

1. Glucose is converted into glucose-6-phosphate (Glc-6-P) in presence of hexokinase (in all tissues) or glucokinase (in liver).
2. Glc-6-P is impermeable to cell membrane and is converted in to Fructose-6-P in presence of phosphohexoseisomerase.
3. Fructose-6-phosphate is phosphorylated to fructose 1,6-bisphosphate by phosphofructokinase (PFK) in presence of ATP. This is an irreversible and a regulatory step in glycolysis.
4. Fructose 1,6-bisphosphate splits into two 3-carbon compounds: glyceraldehyde 3-phosphate and dihydroxyacetone phosphate by the enzyme aldolase.
5. Dihydroxyacetone phosphate and glyceraldehyde 3-phosphate are inter-converted into each other in presence of triose phosphate isomerase. Thus, two molecules of glyceraldehyde 3-phosphate are obtained from one molecule of glucose.

Energy generation phase or payoff phase or yielding phase

Payoff phase yields 4 ATP, 2 NADH and 2 pyruvate molecules per glucose molecule. The steps are:

6. Glyceraldehyde 3-phosphate is converted into 1,3-bisphosphoglycerate along with 2 molecules of NADH in presence of dehydrogenase enzyme. NADH is energy rich compound and passes in to electron transport chain (ETC) under aerobic conditions and produce ATP molecules (1 NADH= 3 ATP).
7. 1,3-bisphosphoglycerate is further converted into 3-phosphoglycerate in presence of phosphoglycerate kinase and 2 ATP molecules are produced.
8. 3-phosphoglycerate is converted in to 2-phosphoglycerate by phosphoglyceratemutase.

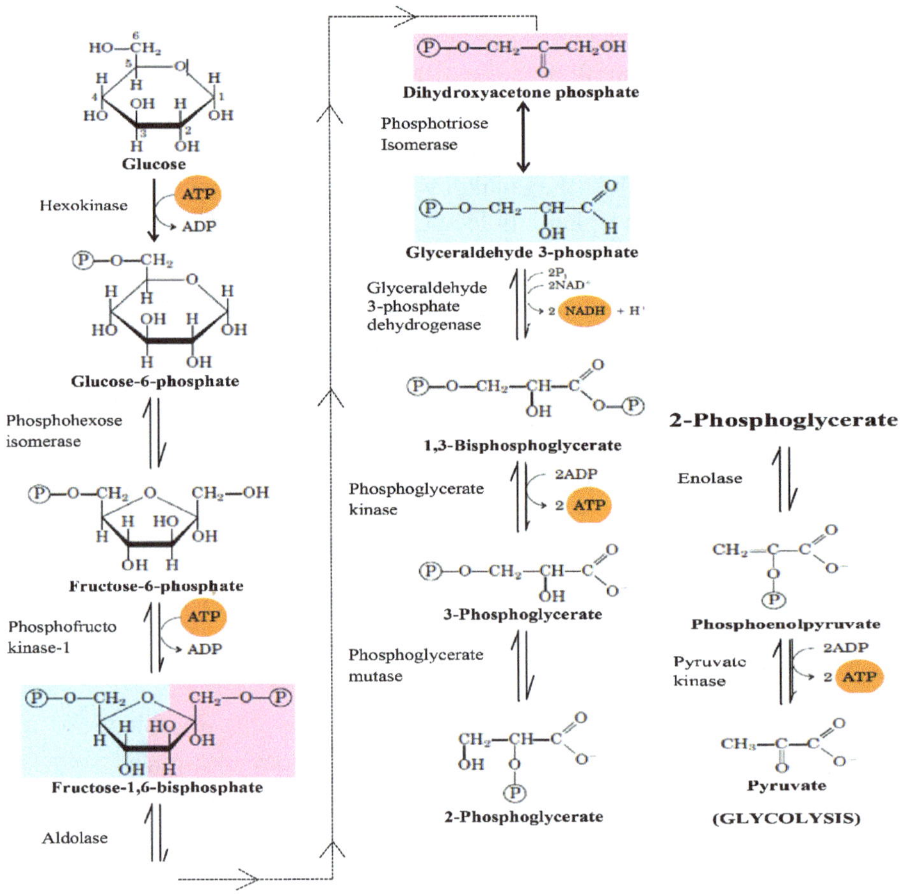

Fig. 5.1: Reactions of glycolysis

9. 2-phosphoglycerate is further catalyzed to phosphoenol pyruvate (PEP) in the presence of enolase. Enolase requires Mg^{2+}or Mn^{2+}ions for enzymatic activity and is inhibited by F^-ions.
10. Pyruvate kinase converts PEP into end product of glycolysis i.e. pyruvate and produce 2 ATP molecules.

It is to be noted that under standard intracellular conditions, glycolysis is essentially irreversible process. Overall, glycolysis can be shown by the following equation:

$$\text{Glucose} + 2\text{NAD}^+ + 2\text{ADP} + 2\text{Pi} \longrightarrow 2\text{ Pyruvate} + 2\text{NADH} + 2\text{H}^+ + 2\text{ATP} + 2\text{H}_2\text{O}$$

In this way, it can be observed that pyruvate is the end product of glycolysis, which further, can be converted in to lactate under anaerobic conditions or acetyl-CoA under aerobic conditions. The fate of pyruvate is shown in Fig. 5.2.

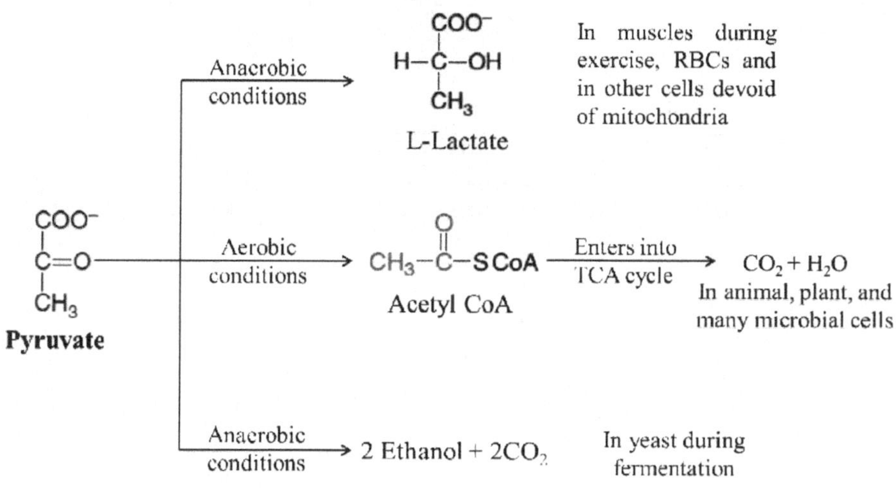

Fig. 5.2: Fate of pyruvate under different conditions

Total ATP production during glycolysis

Under aerobic conditions, one molecule of NADH is converted into 3 ATPs during electron transport chain (ETC); and thus, in total 6 ATP + 2 ATP = 8 ATP are produced. However, under anaerobic conditions, ETC is not possible and NADH is not converted into ATP; consequently, only 2 ATP are generated through glycolysis.

Citric acid cycle

Citric acid cycle also known as tricarboxylic acid (TCA) cycle or Krebs cycle after the name of its discoverer Hans Krebs, is the central metabolic pathway, which not only provides the energy, but also generates several other intermediates which act as precursors forvarious other compounds.This cycle utilizes about 2/3rd of the total oxygen consumed by the body and produce 65-70% of total energy. Citric acid cycle is the final common central pathway for oxidation of carbohydrates, fats and amino acids. Under aerobic conditions, pyruvate (formed from glucose in glycolysis) and acetyl CoA (from fats and amino acids) are ultimately converted into CO_2 and H_2O via TCA pathway. This aerobic phase of catabolism is known as 'Respiration'. All the enzymes participating in TCA cycle remain present in mitochondria, thus, TCA pathway is strictly aerobic in nature, unlike glycolysis which can be operated in both aerobic as well as anaerobic conditions.

Reactions of TCA cycle

Pyruvate formed in glycolysis is converted into acetyl CoA in presence of pyruvate dehydrogenase, which is the starting point of citric acid cycle. This conversion along with involvement of amino acids and fatty acids is shown below (Fig. 5.3).

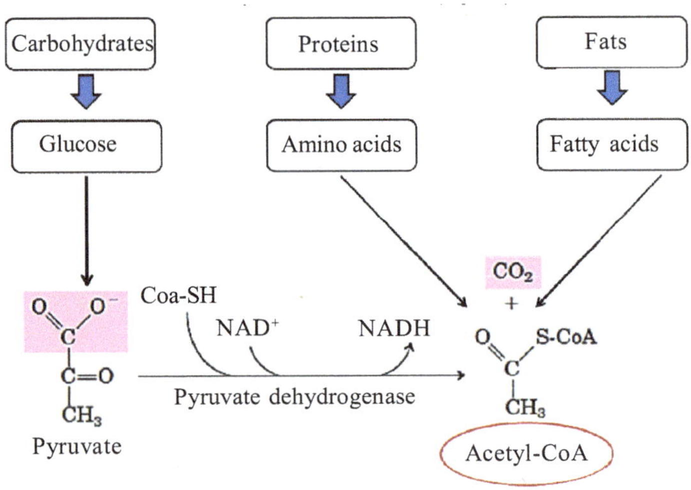

Fig. 5.3: Conversion of pyruvate and other molecules into acetyl CoA

Steps of TCA cycle

With the help of citrate synthase, acetyl CoA combines with oxaloacetate and forms **citrate**, a tricarboxylic acid and the first intermediate. That's why Krebs cycle is also known as 'Tricarboxylic acid cycle'. The reactions of TCA cycle are given in Fig. 5.4.

1. Citrate is first isomerized to cis-aconitate which is an intermediary compound followed by isocitrate in presence of aconitase.

2. Isocitrate is converted to a-ketoglutarate with an intermediary compound oxalosuccinate in presence of isocitrate dehydrogenase. Two molecules ofNADH and CO_2 each (per 2 molecules of acetyl CoA) are liberated in this step.

3. Through oxidative decarboxylation, α-ketoglutarate is catalyzed to succinyl-CoA with the help of a-ketoglutarate dehydrogenase along with 2NADH and $2CO_2$ molecules.

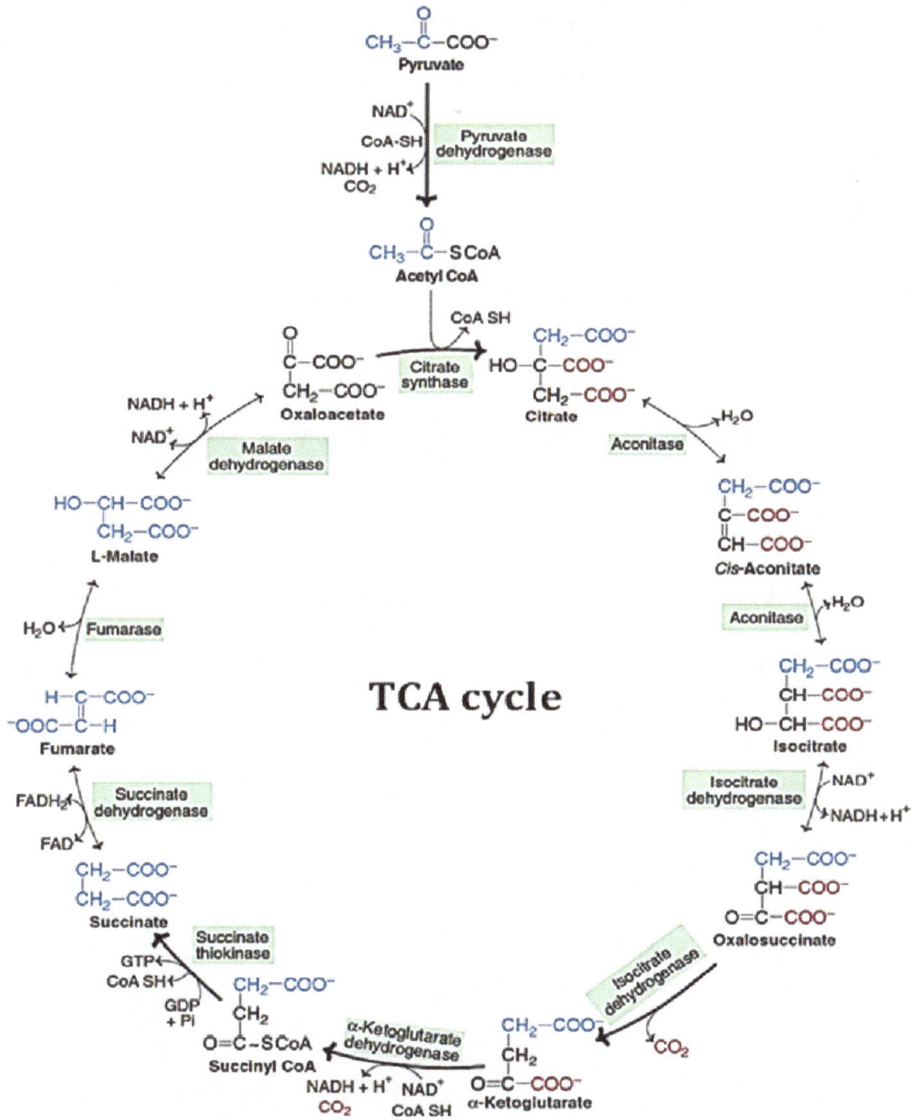

Fig. 5.4.: Reactions of tricarboxylic acid (TCA) cycle

4. Succinyl-CoA is converted to succinate by succinate thiokinase with generation of 2 GTP molecules. GTP can easily be converted into ATP by nucleoside diphosphate kinase enzyme.
5. Now, succinate is converted to fumerate with generation of $FADH_2$. (1 $FADH_2$= 2 ATP)
6. In this hydration step, fumerase catalyze the conversion of fumerate to malate, which finally oxidized to oxaloacetate by malate dehydrogenase. At this stage, 2 NADH are produced.

Overall, TCA cycle can be summarized by the following equation:

$$2\text{ Acetyl CoA} + 6\text{ NAD}^+ + 2\text{ FAD}^+ + 2\text{ GDP} + 2\text{ Pi} + 4\text{ H}_2\text{O} \longrightarrow 4\text{CO}_2 + 6\text{ NADH} + 6\text{H}^+ + 2\text{ FADH}_2 + 2\text{ GTP} + 2\text{CoA}$$

Total number of ATPs formed:

Glucose ⟶ 2 Pyruvate (Glycolysis)	2 ATP + 2 NADH = 8 ATP
2 Pyruvate ⟶ 2 Acetyl CoA	2 NADH* = 6 ATP
2 Acetyl CoA ⟶ CO_2 + H_2O (TCA cycle)	2 GTP + 6 NADH + 2 $FADH_2$*= 24 ATP
Total ATP for **Glucose** ⟶ **CO_2 + H_2O**	**38 ATP** (under aerobic conditions)

*1 NADH = 3 ATP; 1 $FADH_2$= 2 ATP

Amphibolic nature of TCA cycle

During TCA cycle, several intermediate metabolites are formed which act as precursor of various non-essential amino acids, fatty acids, sterols, etc. In other words, TCA cycle is catabolic as well as anabolic; thus known as amphibolic in nature. Amphibolic nature of TCA cycle is shown below (Fig: 5.6)

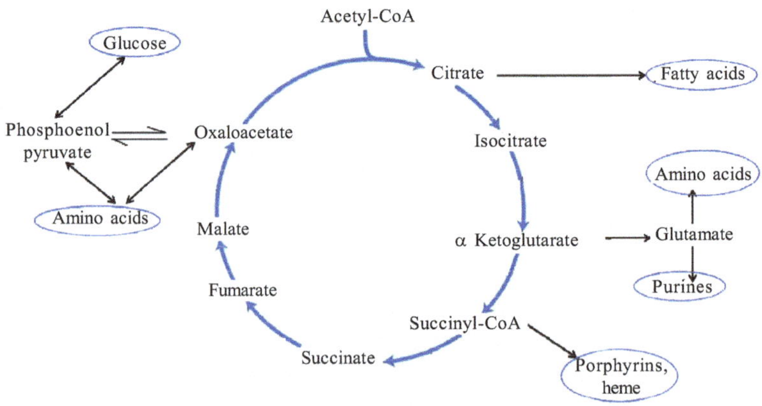

Fig. 5.6: Amphibolic nature of TCA cycle

Oxidative phosphorylation

As discussed above, one molecule of glucose gives 38 ATPs, out of which 2 ATPs are directly formed during glycolysis and 2 ATPs (GTP) are formed in TCA cycle. Rest 34 ATPs are formed from reduced electron carriers (NADH and $FADH_2$) formed in glycolysis & TCA cycle. This reaction occurs in inner mitochondrial membrane, where four complex proteins (I-IV), ubiquinone (Q) and ATP synthase are involved (Fig. 5.7). NADH and $FADH_2$ are oxidized by complex I and II, respectively, and produce NAD^+ and FAD resulting in liberation of free electrons (e^-) and protons (H^+). These free electrons are transferred to mobile electron carrier called ubiquinones (Q), which are further transported to complex III and IV proteins embedded in mitochondrial membrane. In this process, released energy is captured by protons which are pumped out of the matrix to inter-membrane space resulting in generation of a proton gradient with higher protons concentration on outer side and lower concentration at inner side (matrix). This gradient represents a stored form of energy. In presence of electrons, protons (in matrix) react with oxygen and form water molecules. This series of electron transport via protein complexes is known as **electron transport chain** (ETC). Another enzyme of membrane i.e. ATP-synthase utilizes the energy of pre-set up proton gradient and generates ATP from ADP and this process is known as **chemiosmosis**. ETC and chemiosmosis, together are known as **oxidative phosphorylation**.

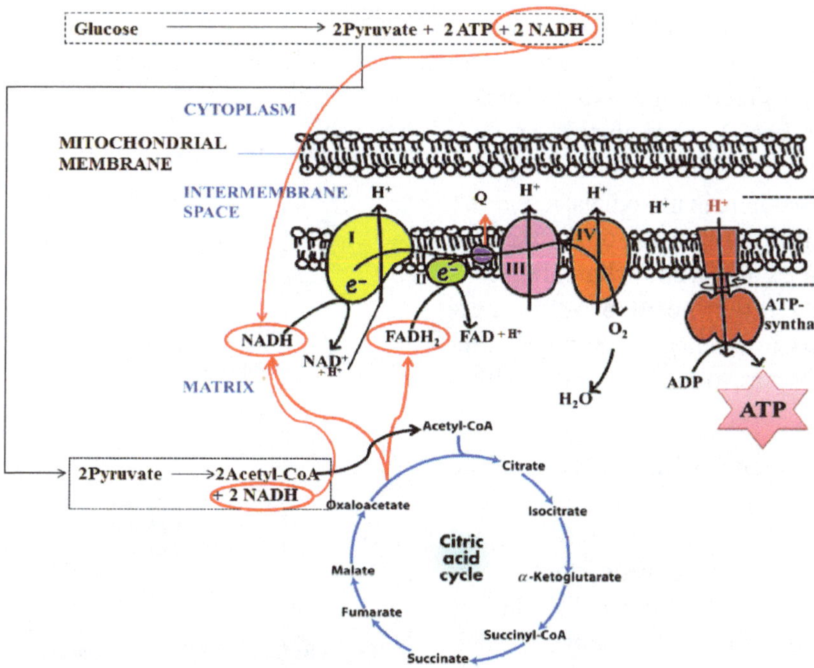

Fig. 5.7: Oxidative phosphorylation of NADH and $FADH_2$

In this way, all reduced forms of energy (NADH, $FADH_2$) formed during glycolysis and TCA cycle, are oxidized during oxidative phosphorylation and generates ATP. Overall net reaction of glucose conversion to ATP can be shown as below:

$$C_6H_{12}O_6 + 6O_2 + 38\ ADP + 38\ Pi \longrightarrow 6CO_2 + 6H_2O + 38ATP$$

Gluconeogenesis

In mammals, some tissues and organs such as brain, nervous tissue, renal medulla, RBCs, testis and embryonic tissues solely depend on the glucose for energy. They cannot derive their energy requirements from other sources like fatty acids or amino acids. Sometimes, during exercise or between meals or fasting stage, sufficient amount of blood glucose is not available for utilization. In such situations, other anabolic methods are needed which could synthesize glucose/carbohydrates from non-carbohydrate sources such as amino acids, fatty acids, lactate, pyruvate, glycerol, etc. This phenomenon of synthesizing glucose from non-carbohydrate sources is known as gluconeogenesis (*glucose-new-synthesis*). Though this process occurs in all animals, plants, fungi and microorganism; in mammals, it majorly occurs in liver and to some extent in renal cortex and in epithelial lining cells of small intestine also. After new synthesis in liver, glucose enters into blood stream to be supplied to other tissues.

Gluconeogenesis can be considered opposite to glycolysis in reactions to some extent. As three steps are irreversible in glycolysis, alternate of those reactions are shown here in Fig. 5.8 along with other reactions.

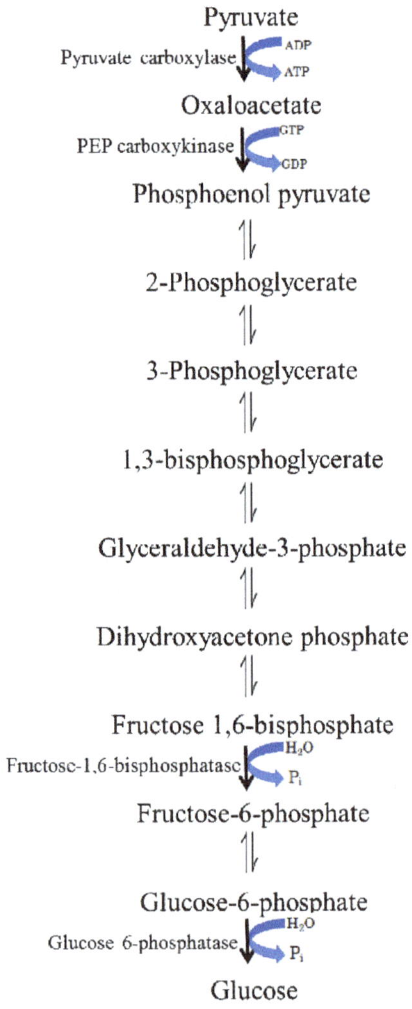

Fig. 5.8: Steps of Gluconeogenesis

Glycogen Metabolism

In organisms, excess glucose is converted into polymeric storage form i.e. glycogen in vertebrates and microorganisms; and starch in plants. In vertebrates, glycogen is majorly present in liver (10% of total liver wt.) and muscles (1-2% of total muscles wt.). Glycogen in muscles provide quick source of energy and can be depleted in <1 hour during exercise and vigorous activity. However, liver glycogen acts as source of glucose (energy) for rest of the organs during fasting or between the meals or when dietary glucose is not available. For example, nerves, brain, retina, etc. cannot use fatty acids or other molecules for energy in any situation; they do require glucose for energy. In such conditions, liver glycogen is broken down to glucose by the process known as 'Glycogenolysis'. In this section, two processes will be discussed:

1. Glycogenesis (synthesis of glycogen from glucose)
2. Glycogenolysis (breakdown of glycogen to glucose)

Glycogenesis

Glycogenesis is simply the synthesis of glycogen from excess glucose in body. It occurs in cytosol and is an energy (ATP) consuming process. The steps of glycogenesis are shown in Fig. 5.9 and details are given below:

1. ***Synthesis of UDP-glucose:*** First glucose is converted into glucose 6-phosphate in presence of glucokinase/hexokinase, which is further converted into glucose 1-phosphate. In presence of UDP-glucose pyrophosphorylase, it is converted into uridinediphosphate glucose (UDP-glucose).
2. ***Initiation of glycogen formation:*** A small glycogen molecule (glycogenin) (n) is required to initiate glycogenesis, in which it acts as a primer and receives one glucose molecule from UDP-glucose in presence of glycogen initiator synthase and forms glycogen molecule (n+1). In this way, this process continues and chain length of glycogen molecule increases.

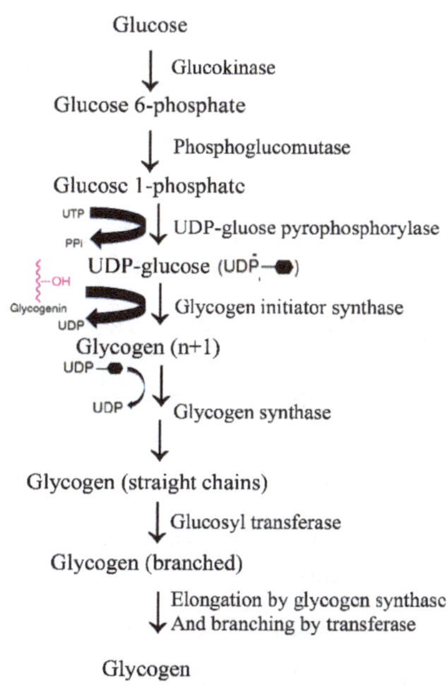

Fig. 5.9: Reactions of Glycogenesis

3. ***Branching of glycogen:*** In presence of glucosyl (α-4-6) transferase (amylo α 1,4 → 6 transglucosidase), branching occurs and branched glycogen molecule is formed in cytosol. In this way, glycogen is further elongated and branched in presence of respective enzymes and glycogen, if formed.

Glycogenolysis

Breakdown of glycogen in liver and muscles at the time of requirement is known as Glycogenolysis. It is not just reversible of glycogenesis. It involves following steps (Fig. 5.10):

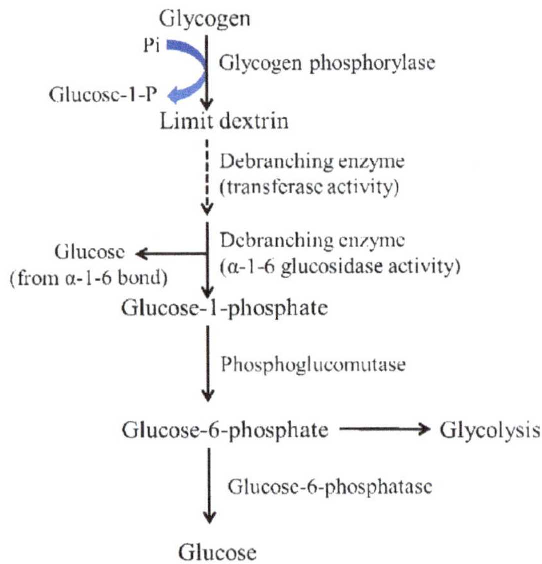

Fig. 5.10: Reactions of Glycogenolysis

1. Glycogen phosphorylase cleaves the glycogen molecule from non-reducing end and release glucose-1-phosphate. This process is known as phosphorolysis and continues till 4 molecules of glucose are left out on branching point of the chain. The resultant molecule is known as limit dextrin.

2. Now debranching enzyme (transferase) transfers the 3 or 4 glucose residue from one position to another so that branching is reduced.

3. Now another debranching enzyme (a-1-6 amyloglucosidase) cleaves at a-1-6 position and releases free glucose molecule yielding straight chain of glucose residues.

4. Phosphorylase further breaks resultant straight chain of glucose and releases glucose-1-P.
5. In presence of phosphoglucomutase, glucose-1-P is converted into glucose-6-P, which may enter into glycolysis pathway or converted into glucose in presence of glucose-6-phosphatase.

Sample questions

1. Define Glycolysis along with different steps.
2. Give fates of pyruvate under aerobic and anaerobic conditions in brief.
3. Give overall reaction of glycolysis and show the number of ATPs formed in different pathways during complete oxidation of glucose molecule.
4. In diabetic patients, how glucose is transferred from blood into the brain cells?
5. What do you understand by amphibolic nature of TCA cycle?
6. Why Krebs cycle is also known as TCA cycle? Give its reactions.
7. What do you understand by oxidative phosphorylation?
8. Differentiate between glycolysis and TCA cycle.
9. Compare Glycogenesis, Glycogenolysis and Gluconeogenesis.
10. Define Gluconeogenesis. Discuss its significance in human body.

Suggested Readings

Maughan, R., 2005. Basic metabolism II: carbohydrate. Surgery (Oxford), 23(5), 154-158.

Nelson, D.L., Lehninger, A.L. and Cox, M.M., 2008. Lehninger Principles of Biochemistry. Macmillan.

Satyanarayana, U. and Chakrapani, U., 2006. Biochemistry, Books and allied (P) Ltd. Reed Elsevier India Private Limited.

http://www.biologydiscussion.com/metabolism/carbohydrates-metabolism/metabolism-of-carbohydrates-10-cycles-with-diagram/11242

https://www.biochemden.com/carbohydrate-metabolism-overview/

6

Lipid Metabolism

Introduction

The diverse cluster of naturally occurring organic compounds including fats, oils, fatty acids, waxes, phospholipids, eicosanoids, terpenes, steroids and fat soluble vitamins that do not prefer to interact with water but soluble in non-polar organic solvents like ether, chloroform, benzene, acetone, are called as lipids. The main functions of lipids in biological system include storage and transport of metabolic fuel, structural component of cell membranes, insulation & protection of vital organs from shock and carrier of fat soluble vitamins. Contradictory to proteins, lipids are not polymers in nature and are hydrophobic due to presence of a long hydrophobic chain. Apart from this, lipids also act as precursor molecules for several pigments, hormones and signaling molecules.

Classification of lipids

The lipids may be classified into simple, compound and derived lipid as shown in the Fig. 6.1.

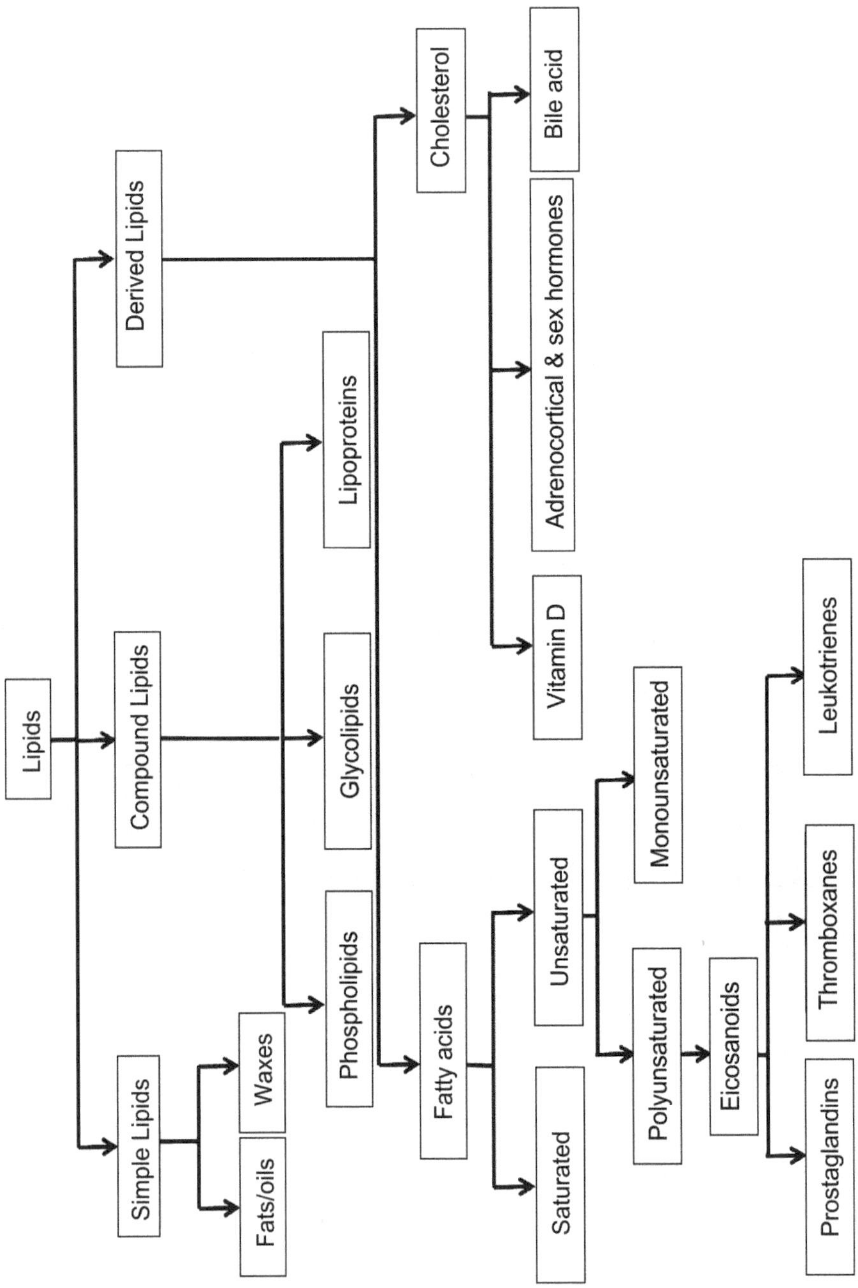

Fig. 6.1: Classification of Lipids

(A) Simple Lipids

Simple lipids or Homo lipids are the esters of fatty acids with various alcohols.These include simple & mixed triglycerides and waxes. The lipids which are solid at room temperature are called fats as they contain a large proportion of saturated fatty acids. Lipids which are liquid at room temperature are called oils as their unsaturated fatty acid content is relatively high.Triglycerides (also called as triacylglycerol) are triesters of fatty acids with glycerol as shown in the Fig. 6.2. If the three fatty acids attached are same, then it is called as simple triglyceride and if two or more fatty acids of different types are attached, it is called mixed triacylglycerol. Most of the fat in body is stored in the form of triglycerides. They are also called as 'neutral' fats as no charge is present on them.

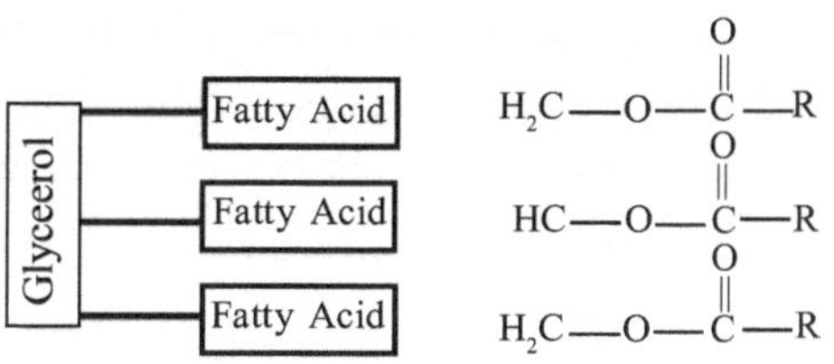

Fig. 6.2: General structure of triglycerides

Waxes, a second group of neural lipids are also classified under simple lipids. The formation of waxes takes place by esterification of long chain fatty acids with higher alcohols, normally having 14 to 36 carbon atoms. They are strongly hydrophobic due to presence of a long hydrocarbon chain. This hydrophobic nature allows their use as water repellent on leaves of some plants, feathers of birds and cuticles of some insects.

(B) Compound or complex lipids

These are the lipids in which along with fatty acids and alcohols, other groups like phosphates, nitrogenous base, proteins and carbohydrates are also attached. They can be further classified into phospholipids, glycolipids and lipoproteins based on the group attached.

(a) Phospholipids

Phospholipids are those complex compounds in which along with fatty acids and alcohols, a phosphoric acid molecule is also present as nitrogenous base. They are amphipathic molecules and based on the type of alcohol present they may be further classified as glycerophospholipids or phosphoglycerides and sphingophospholipids or phosphosphingolipids.

Glycerophospholipids are the most frequently found phospholipids in cell membranes. They are made up of two fatty acids esterified with a glycerol backbone, a phosphoric acid molecule and an alcohol.

Phosphatidic acid, the simplest glycerophospholipid is the precursor of all higher glycerophospholipids. The basic structure of phosphatidic acid is as shown in the Fig. 6.3

The different derivatives of phosphatidic acid are formed by replacing 'X' with different alcohols in the structure.

If X = H: Phosphatidic acid

X = Choline: Phosphatidylcholine or lecithin

X = Ethanolamine: Phosphatidylethanolamine or cephalin

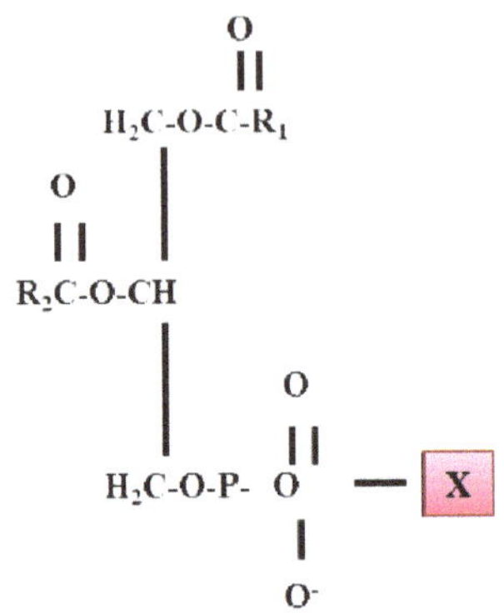

Fig. 6.3: Structure of phosphatidic acid

X = Serine: Phosphatidylserine

X = Glycerol: Phosphatidylglycerol

X = Inositol: Phosphatidylinositol

X = Phosphatidylglycerol: Diphosphatidylglycerol

In phosphoglycerides, out of the two fatty acids attached with the glycerol backbone, the first is saturated and the other one is unsaturated.

Sphingophospholipids contains a long chain amino alcohol called as sphingosine, instead of glycerol.

Ceramide is the simplest sphingolipid molecule and precursor or structural parent of all complex sphingolipids-formed by replacing the hydroxyl group with other groups as represented by letter 'X' in the Fig. 6.4

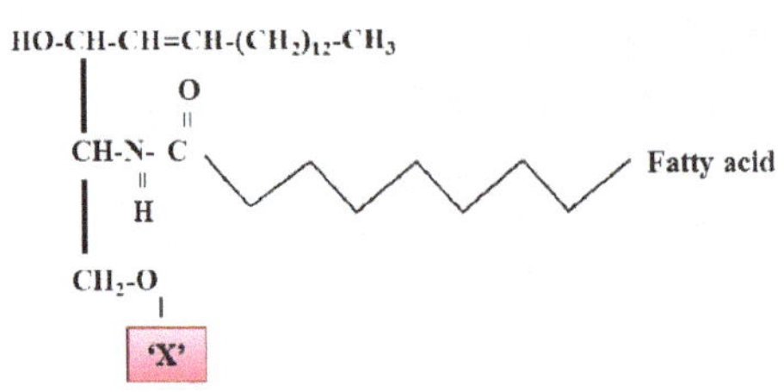

Fig. 6.4: Structure of ceramide

Phospholipid synthesis: As discussed earlier in this chapter, the phospholipids are derived from the compound called as phosphatidic acid. Phosphatidic acid is basically a diacylglycerol (DAG) with a phosphate group at 3rd carbon.

Both **phosphatidyl choline (PC) and phosphatidyl ethanolamine (PE)** are most abundant phospholipids found in biological membrane of eukaryotic organisms. The main route of their synthesis starts with choline and ethanolamine obtained either through diet or from preexisting phospholipids. Choline and ethanolamine after phosphorylation combines with cytidine triphosphate (CTP) in presence of enzyme choline kinase and ethanolamine kinase to form cytidinediphosphate (CDP) choline and CDP ethanolamine, respectively. CDP choline, combine with 1, 2 diacylglycerol, in presence of enzyme phosphocholine diacylglycerol transferase to form phosphatidyl choline or lecithin. Similarly, CDP ethanolamine, combine with 1, 2 diacylglycerol in presence of enzyme phosphoethanolamine diacylglycerol transferase to form phosphatidyl ethanolamine or cephalin as shown in the Fig. 6.5 & 6.6.

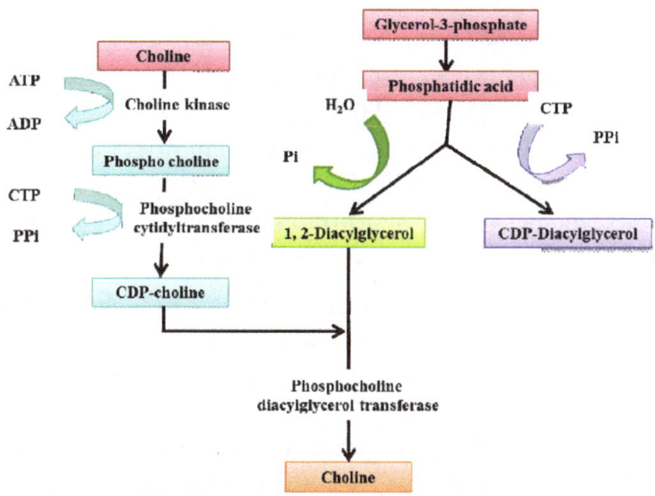

Fig. 6.5: Biosynthesis of choline

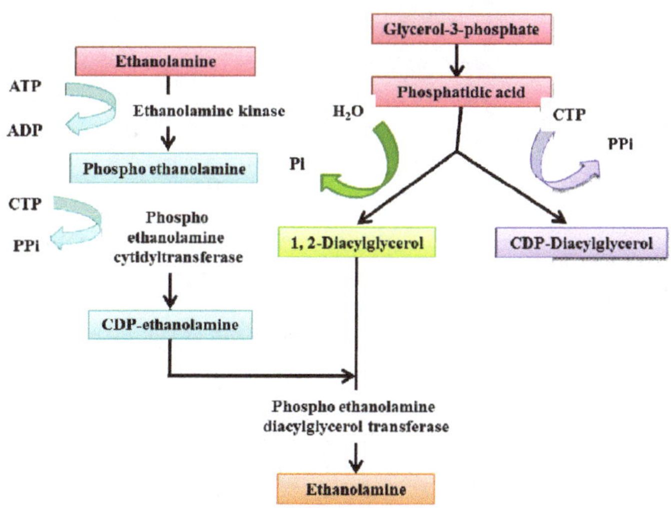

Fig. 6.6: Biosynthesis of ethanolamine

Phosphatidyl serine (PS) is synthesized from PE (Fig-6.7) by exchange of its ethanolamine group with free serine residue.

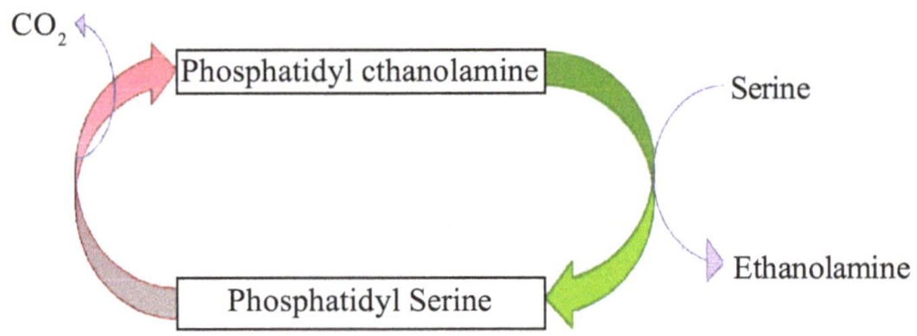

Fig. 6.7: Biosynthesis of Phosphatidyl serine

Phosphatidyl inositol (PI) is the unique phospholipid as stearic and archadonic acids are the most commonly found fatty acids at 1st and 2nd carbon atom. First, CDP-Diacylglycerol is synthesized from phosphatidic acid and the reaction is catalyzed by the enzyme CTP-phosphate cytidylyltransferase. CDP-Diacylglycerol, combines with inositol in presence of enzyme CDP-Diacylglycerol inositol transferase to form PI as shown in Fig. 6.8.

Phosphatidyl glycerol (PG) is abundantly found in mitochondrial membranes and act as a precursor of cardiolipin. **Cardiolipin** is made up of two phosphatidic acid molecules connected by a molecule of glycerol. CDP-Diacylglycerol synthesized from phosphatidic acid combines with glycerol-3-phosphate to form phosphatidyl-glycerol-3-phosphate as an intermediate compound, ultimately synthesizing PG. Two molecules of PG combines together to formcardioliolipin as shown in Fig. 6.8.

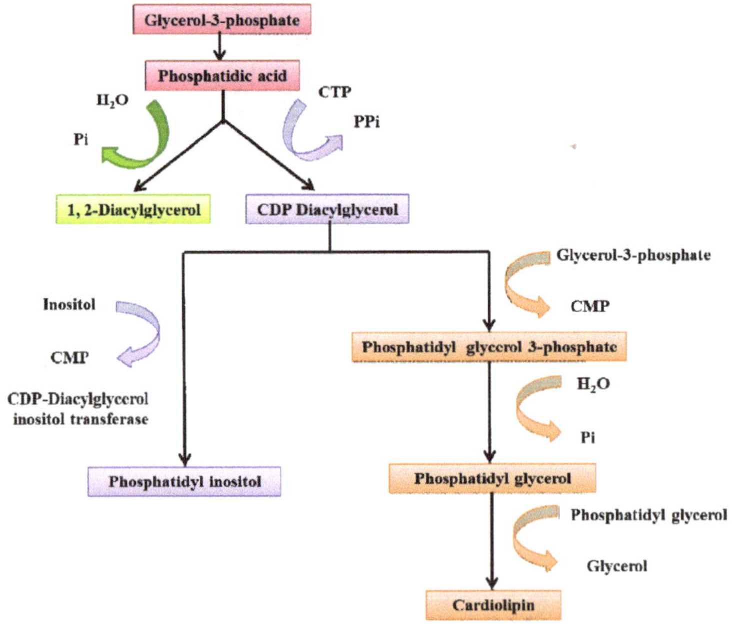

Fig. 6.8: Biosynthesis of phosphatidyl inositol and cardiolipin

Plasmalogens are the phospholipids in which the fatty acid at 1st carbon atom is attached with ether linkage instead of ester linkage. They are a group of several structurally related compounds, playing important role in blood platelet aggregation, therefore, also known as platelet activation factor (PAF). The major classes of plasmalogens include phosphatidylcholines, phosphatidylethanolamines and phosphatidylserines. Peroxisomes and endoplasmic reticulum are the preferred locations of enzymes accountable for plasmalogen biosynthesis. First of all, dihydroxyacetone phosphate (DHAP) is acylated at Sn-1 position to form 1-acyl DHAP, with the help of an enzyme known as dihydroxy acetone phosphate acyl transferase (GNPAT). At the same time one molecule of fatty acyl CoA is reduced to fatty alcohol. This NADPH dependent reduction is accomplished in presence of two enzymes known as fatty acyl CoA reductase-1 & fatty acyl CoA reductase -2. This fatty alcohol then enters into the peroxisomes, where it displaces the fatty acid attached at Sn-1 position of 1-acyl DHAP, resulting in formation of 1-alkyl DHAP. This step of plasmalogen biosynthesis is catalyzed by an enzyme known as alkyl dihydroxy acetone phosphate synthase (AGPS) as shown in Fig. 6.9. Finally, 1-alkyl DHAP goes through a sequence of reactions to get converted into plasmalogens.

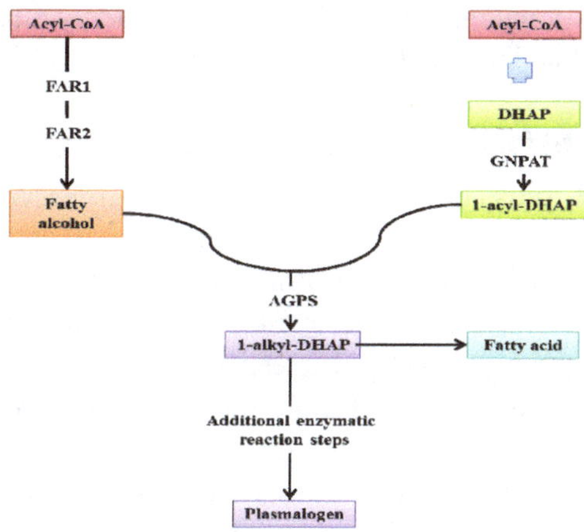

Fig. 6.9: Biosynthesis of plasmalogens

Sphingomyelin is one of the main structural components of membranes of nervous tissues. This class of phospholipids differs from other phospholipids in having the sphingosine rather than glycerol as the alcohol base. Palmitoyl-CoA and serine combines together and undergo a sequence of reactions to form sphingosine. The latter is then acylated to form ceramide. Ceramide combines with CDP-choline to synthesize sphingomyelin as shown in Fig. 6.10.

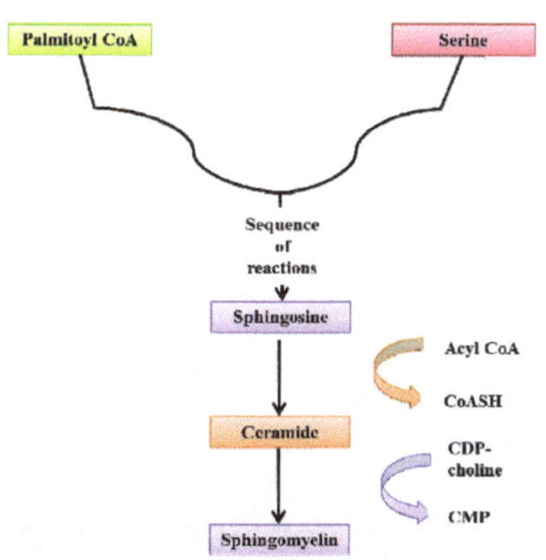

Fig. 6.10: Biosynthesis of Sphingomyelin

Degradation of phospholipids: The phospholipids are degraded by phospholipases found in all tissues and pancreatic juice. Various types of toxins, venoms and pathogenic bacteria also possess phospholipase activity. They utilize this activity to dissolve cell membranes and spread the infection.

The fatty acid present at first carbon atom in phospholipids is cleaved by phospholipase A_1, resulting in formation of lysophospholipids. Further, the fatty acid from 2nd carbon atom of lysophospholipids is removed by two enzymes namely, lysophospholipase and phospholipase B. In phosphatidyl inositol, the fatty acid at 2nd carbon atom is also cleaved by phospholipase A_2 and the arachidonic acid liberated act as a substrate for prostaglandin synthesis. The bond between phosphate and glycerol backbone is cleaved by phospholipase C. However, phospholipase D specifically removes the nitrogen base. The enzyme sphingomyelinase degrades sphingomyelins into phosphorylcholine and ceramide. The latter is further hydrolyzed into sphingosine and free fatty acids.

(b) Glycolipids

Lipids containing carbohydrates or saccharides as nitrogenous base are called as glycolipids. The carbohydrates attached may be glucose or galactose molecules. Depending upon the presence of glycerol or sphingosine as the backbone of the structure, they are classified as glycosphingolipids and glycoglycerolipids, respectively. Fig. 6.11 represents the general structure of these two glycolipids. Based on the number of sugar molecules attached with ceramide, they are further classified as cerebroside (single sugar molecule attached), globosides (two or more than two sugar molecules attached) and gangliosides (complex oligosaccharide attached).

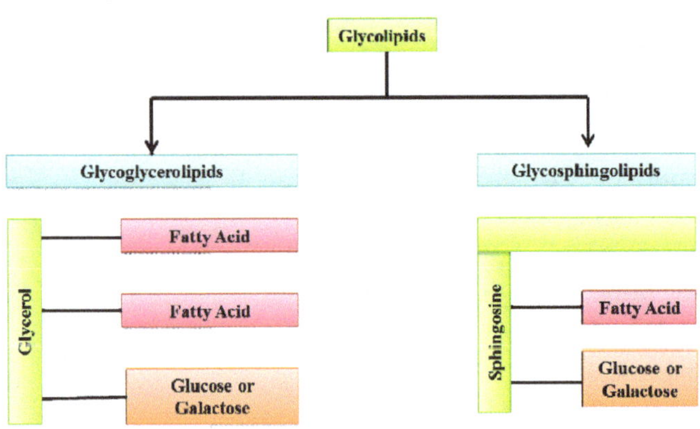

Fig. 6.11: Classification of glycolipids

Glycolipid metabolism: Synthesis of cerebrosides occurs in Golgi apparatus by stepwise transfer of glycosyl monomers to the acceptor molecule (ceramide) in presence of glycosyltransferases, specific for different sugars. Degradation of glycolipids occurs through the help of enzymes present in the lysosomes. These enzymes fuse with endocytotic vesicles and starts cleaving specific bonds of glycolipids. This process proceeds as per the "Last on, First off" rule; means the group which was added in the last during the synthesis, gets removed first during the degradation process. Both synthesis and degradation processes runs in a balanced way in a normal individual. However, if a lysosomal enzyme, used for hydrolytic degradation of glycolipids is partially or completely missing due to some inherited disorder, then that sphingolipid starts accumulating in the body, leading to diseases called sphingolipidoses.

(c) Lipoproteins

Lipoproteins are macromolecular complexes of lipids with proteins. They consist of a hydrophobic core of neutral lipids (triglycerides and cholesterol esters) and an outer relatively hydrophilic coating of phospholipids. Various types of associated proteins called as apolipoproteins and cholesterol molecules are rooted in outer coating. Lipoproteins act as circulatory molecules. Most of the lipid molecules are insoluble in blood plasma; therefore, lipoproteins act as carriers of triacylglycerols, phospholipids, cholesterol and cholesterol esters. Depending upon the difference in density, the four main categories of lipoproteins comprise of chylomicrons, very low density lipoproteins (VLDL), low density lipoprotein (LDL) and high density lipoprotein (HDL). The important characteristics of these major classes are as shown in Table 6.1.Chylomicrons, the lowest density lipoproteins are involved in transport of exogenous lipids from intestine to the cells while VLDL transports endogenous lipids from liver to different cells. LDL involved in transport to cholesterol to the different cells and organs from liver is regarded as 'Bad cholesterol' and HDL involved in reverse transport of cholesterol from cells to liver is regarded as 'Good cholesterol'.

Table 6.1: Characteristics of major lipids

Lipoprotein	Density (g/ml)	(% by mass)				Apolipoprotein
		PRO	CHO	TG	PL	
Chylomicrons	< 1.006	1	4	90	5	A, B, C, E
VLDL	0.95-1.006	8	25	55	12	B, C, E
LDL	1.006-1.063	20	55	5	20	B
HDL	1.063-1.210	50	20	5	25	A, C, E

PRO, protein; CHO, cholesterol; TG, triglyceride; PL, phospholipids

(C) Derived lipids

Derived lipids are the derivatives produced by hydrolysis of different simple and compound lipids. These include eicosanoids, fatty acids, cholesterol, fat soluble vitamins and sex hormones.

Eicosanoid

These are a group of very potent biological signaling molecules formed by either enzymatic or non-enzymatic oxidation of PUFA having 20 carbon fatty acids. Eicosanoids may be further classified as prostaglandins, thromboxane, prostacyclin and leukotriene.

Prostaglandins, a group of hormone like compounds were first discovered by Noble Laureate in human semen. The only difference between prostaglandins and hormones is that, hormones are transported through blood stream to the target organ but prostaglandins are synthesized locally near the site of action. Previously,it was assumed that prostaglandins are synthesized in prostate glands but later it was realized that all body cells except red blood cells can synthesize them. Prostaglandins are derived from hypothetical C_{20} fatty acid called as prostanoic acid (Fig. 6.12). They are responsible for diverse functions like uterine contraction, lowering blood pressure, stimulation of platelet aggregation, mediation of inflammatory response, regulation of nerve transmission, sensation of pain, inhibition of gastric secretions and platelet aggregation.

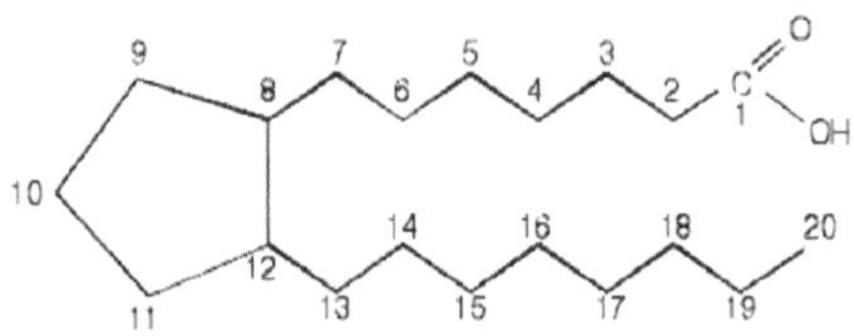

Fig. 6.12: Structure of prostanoic acid

Thromboxanes are also derivative of prostanoic acid with a six member oxane ring instead of five carbon cyclopentane ring as in the case with prostaglandins. They act as vasoconstrictors, play role in platelet aggregation, stimulate steroid production by adrenal cortex and increases mobility of calcium ions from bones.

Leukotrienes are a family of conjugated trienes that do not contain a ring structure and are hydroxyl fatty acid derivatives of arachidonic acid. They are involved in chemotaxis, inflammation and allergic reactions.

Cholesterol

Cholesterol is an amphipathic steroidal molecule with molecular formula $C_{27}H_{45}OH$ and it is also the major sterol present in animal tissues. The structure of the cholesterol may be divided into three parts, a hydrocarbon tail, a hydroxyl group and a ring structure made up with four hydrocarbon rings. These four rings A, B, C & D are arranged in a specific molecular configuration as shown in the Fig. 6.13. The molecule is little bit polar in nature due to the presence of a hydroxyl group.Cholesterol is classified in the category of sterols due to presence of these four fused rings and hydroxyl group. The third part of the structure is hydrocarbon tail. Both the rings region and the tail region are non-polar in nature. Cholesterol is an important component of cell membranes and precursor of several compounds like bile acids, vitamin D and steroid hormones produced by adrenal cortex and gonads.

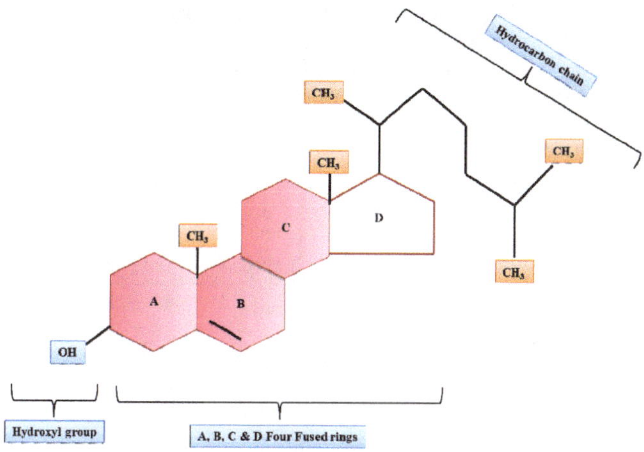

Fig. 6.13: Structure of Cholesterol

Cholesterol biosynthesis: Cholesterol biosynthesis is a complex multistep process and each step is regulated by enzymes depending upon the availability of dietary cholesterol intake. In case of low dietary cholesterol, its biosynthesis is up regulated and in case of excess dietary intake, excretion will increase and its biosynthesis will be down regulated.

Mevalonate formation is the first step of cholesterol biosynthesis. Two molecules of acetyl-CoA condense with each other to form acetoacetyl-CoA in presence of enzyme thiolase. Then, the acetoacetyl-CoA reacts with another molecule of acetyl-CoA in presence of enzyme hydroxyl methyl glutarate-CoA synthase to form 3-hydroxy-3methyl glutaryl-CoA (HMG-CoA). This HMG-CoA is ultimately reduced by HMG-CoA reductase to form mevalonate. The latter is the committed but also rate limiting step of the whole process.

In the subsequent steps, mevalonate is converted into two activated five carbon isoprenoid compounds known as isopentanyl-5-pyrophosphate and dimethylallyl pyrophosphate. These isoprenoid compounds further condensates and polymerizes through a series of reactions to form a 30 carbon molecule, called squalene. Squalene, the precursor for all steroid compounds, is then converted in to a four ring compound, lanosterol. Lanosterol is ultimately transformed into cholesterol through a series of reactions as shown in the Fig. 6.14.

Cholesterol utilization: The production and utilization proportions of cholesterol in a normal healthy individual are 1g/day and 0.3g/day, respectively. Humans are not capable to metabolize the sterol nucleus in cholesterol; rest is converted into bile acids or utilized for steroid hormone synthesis and vitamin D.

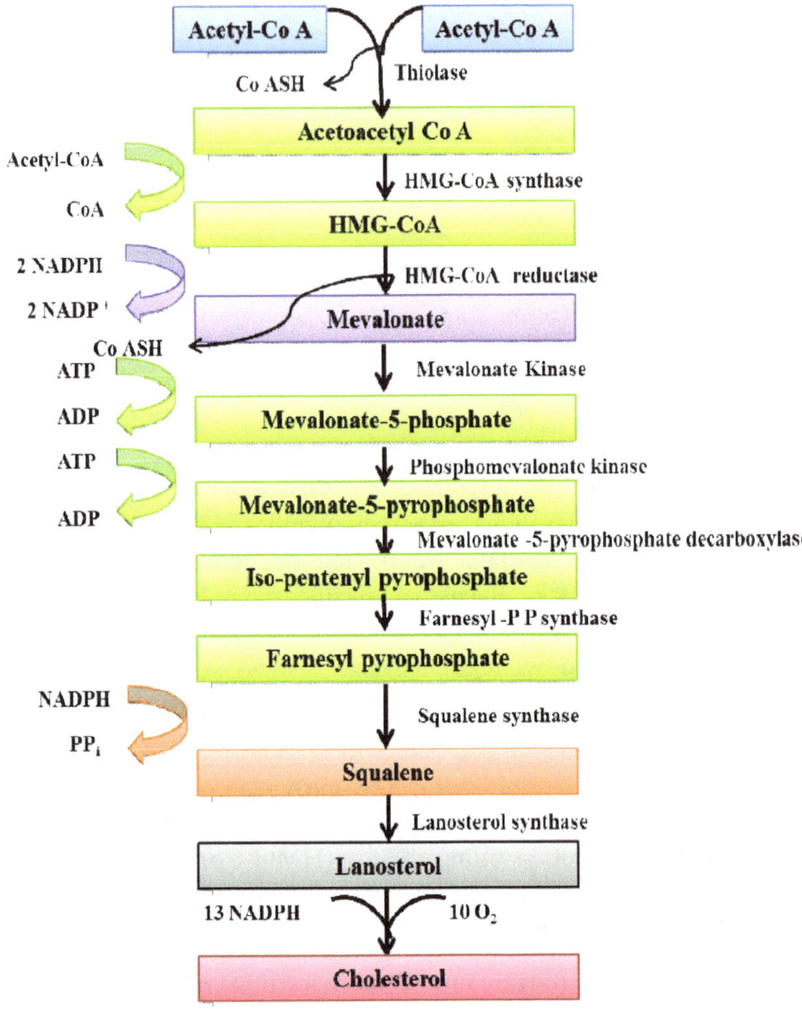

Fig. 6.14: Biosynthesis of cholesterol

1. Bile acid synthesis from cholesterol

Bile acids are the main end product of cholesterol catabolism. They are involved in solubilisation of dietary cholesterol, fat soluble vitamins along with other nutrients and thus promote their absorption and delivery to liver. Liver is also the main organ for complete biosynthesis of bile acids from cholesterol. Bile acid synthesis requires a total of 17 different enzymes which catalyzes different reactions in multiple intracellular compartments such as cytosol, endoplasmic reticulum, mitochondria and peroxisomes. Many of the bile acid metabolites are cytotoxic in nature; therefore, the process of bile acid synthesis is controlled very precisely by genes responsible for encoding of enzymes involved in biosynthesis process.

The first step of the bile acid synthesis pathway involves hydroxylation of the cholesterol at 7th position with the help of enzyme 7-α-hydroxylase (Fig. 6.15). This is the critical and rate limiting step. An alternative pathway is also there, in which hydroxylation of cholesterol at 27th position takes place. Mitochondrial enzyme, called sterol 27 hydroxylase catalyzes the later reaction. The first pathway is referred as neutral pathway, while the later one as acidic pathway. In humans, neutral pathway contributes for synthesis of more than 90% bile acids.

The hydroxyl group at 3rd position in 7-α-cholesterol is epimerized to α-orientation by the enzyme 3 β-hydroxy-δ5-C27-steroid oxidoreductase. Following this reaction, the bile acid intermediates can proceed via two pathways. If the intermediates are acted upon by sterol 12α-hydroxylase enzyme, then cholic acid is formed and if the intermediates escape this step then chenodeoxycholic acid is formed. Both cholic acid and chenodeoxycholic acid are primary bile acids contributing 31% and 45% to the total human bile, respectively.

Before secreting into the lumen, these acids get conjugated with glycine and taurine at the carboxylic end via an amide linkage, yielding glycoconjugates and tauroconjugates, respectively. This conjugation helps in making them more amphipathic and less cytotoxic in nature. In intestine, de-conjugation of primary bile acids occurs, resulting in formation of secondary bile acids (deoxycholic acid and lithocholic acid).

The conjugated bile acids accumulate in the gall bladder and are secreted into the small intestine to act as an emulsifying agent for digestion and absorption of fat and fat soluble vitamins. After their function is over, they are reabsorbed and returned to the liver through portal vein.

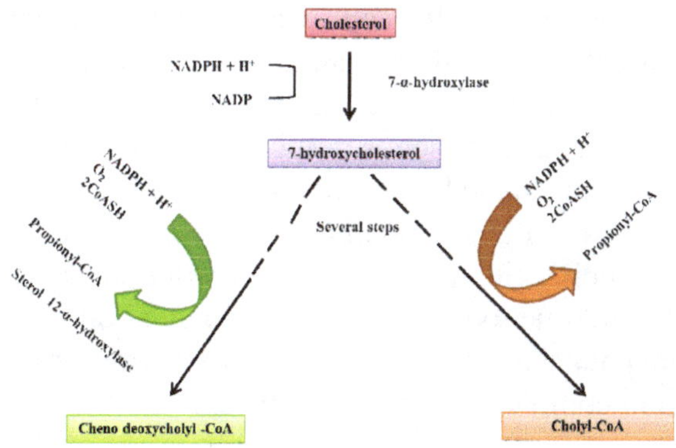

Fig. 6.15: Synthesis of bile acids from cholesterol

Steroid hormone synthesis from cholesterol

Cholesterol acts as a precursor for synthesis of sex hormones such as cortisol, aldosterone, progesterone, testosterone and estradiols. A brief outline of steroid hormone synthesis has been shown in Fig. 616 and more details had already been discussed under Chapter-2 'Hormones'.

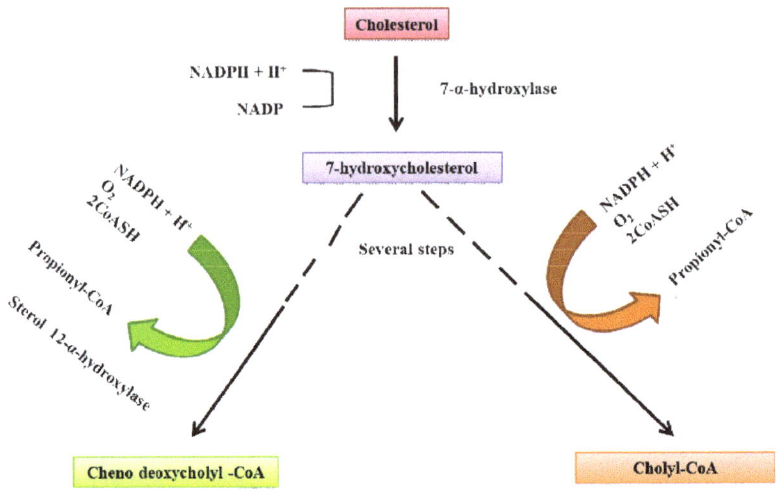

Fig. 6.16: Synthesis of steroid hormones from cholesterol

Vitamin D synthesis from cholesterol

Ultraviolet rays of the sunlight convert 7-dehydrocholesterol (an intermediate in synthesis of cholesterol) into cholecalciferol. The details of the process had already been discussed under Chapter 'Vitamins'.

Fatty acids

Fatty acids are the simplest form and building blocks for complex lipids. They consist of a long hydrophobic hydrocarbon chain and a hydrophilic carboxyl group. Butyric, caproic, caprylic, capric, lauric, myristic, palmitic, stearic and arachidic acids are the predominant naturally synthesized saturated fatty acids. The hydrocarbon chain may be saturated or unsaturated depending upon the absence or presence of double bonds, respectively. Saturated fatty acids possess no double bonds. Unsaturated fatty acids are called as mono and polyunsaturated fatty acids (MUFA and PUFA) based on the presence of one or more than one double bonds in their structure, respectively. In unsaturated fatty acids, the chain becomes rigid due to the presence of double bonds. As a result of that, the chain cannot pack them compactly. Therefore, unsaturated fatty acids have low melting point compared to saturated ones.

Fatty acid metabolism

Metabolism of fatty acids consists of fatty acid synthesis and their degradation for using them as a source of energy. Both these metabolic processes occur in different cell organelles in order to enhance energy utilization. Both the processes are quite simple and almost reverse of each other.

Fatty acid biosynthesis

Cytosol and the chloroplast stroma are the places for fatty acid biosynthesis in animals and plants, respectively. As this process involves addition of two carbon units in single steps, therefore, all naturally occurring fatty acids are having even number of carbon atoms only. These two carbon units are derived from acetyl CoA, which is further obtained from oxidation of glucose. This acetyl CoA is present in the mitochondria of the cell and from there it needs to be transported into the cytosol through the help of acetyl CoA shuttle system (Fig. 6.17). Inside mitochondria, the acetyl CoA reacts with oxaloacetate to form citrate. When the concentration of acetyl CoA and ATP are high inside mitochondrial matrix, the citrate molecule passes out through the mitochondrial membrane into the cytosol, where it reacts with the cytosolic CoA and ATP to dissociate back into acetyl CoA and oxaloacetate. This reaction is catalyzed by the enzyme citrate lyase. The oxaloacetate then gets reduced to malate so as to cross the barrier of mitochondrial membrane and returns back to the mitochondrial matrix.

The two carbon unit (acetyl CoA) gets converted into four carbon unit by a sequence of seven steps of enzyme catalyzed reactions, as shown in the Fig. 6.18.

1. Acetyl CoA is converted into Malonyl CoA. This reaction is catalyzed by enzyme acetyl CoA carboxylase, ATP, bicarbonate ion and biotin.
2. Malonyl CoA is then converted into Malonyl-ACP (Malonyl acyl carrier protein, a two carbon unit).
3. Another molecule of acetyl CoA is converted into Acetyl-ACP.
4. The fourth step in fatty acid biosynthesis is condensation of acetyl-ACP and Malonyl-ACP, synthesized in second and third steps. In this condensation reaction, a four carbon compound, Aceto acetyl-ACP is formed and carbon di oxide is eliminated.
5. Aceto acetyl-ACP is further reduced to D-3-hydroxyacyl-ACP (four carbon unit).
6. D-3-hydroxyacyl-ACP is dehydrated to Crotonyl-ACP (four carbon unit).
7. Crotonyl-ACP is again reduced to Butyryl-ACP.

The reaction steps from 4 to 7 of the sequence are actually the chain elongation steps. These four steps are repeated again and again for synthesis of higher fatty acids and each time the chain gets elongated by addition of two carbon units.

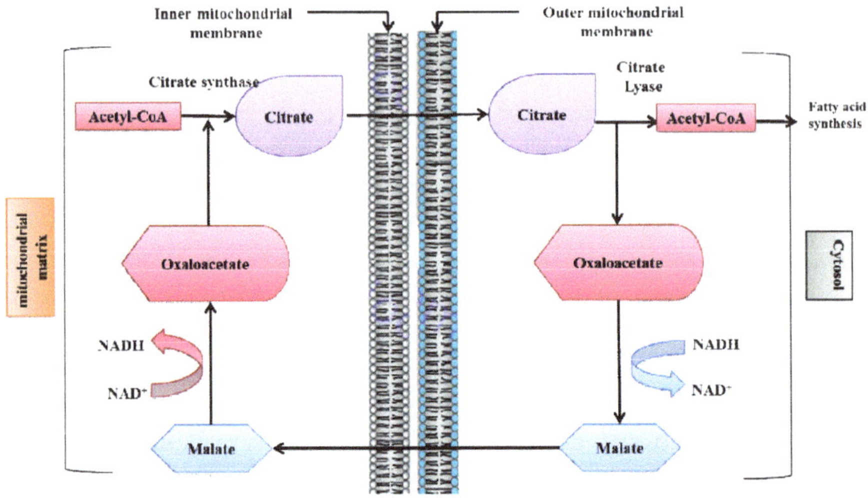

Fig. 6.17: Acetyl CoA shuttle system

Fatty acid oxidation

There are two pathways for fatty acid oxidation.

(1) α-oxidation

(2) β-oxidation

α-oxidation is a minor pathway for fatty acid oxidation, usually opted for degradation of branched chain or odd chain fatty acids. β-oxidation is the most common pathway for fatty acid oxidation.

In this pathway, two carbon fragments are sequentially removed from the carboxylic end of fatty acids. The fatty acids are activated in a reaction with ATP and CoASH to form fatty acyl-CoA. This activation process is catalyzed by an enzyme called acyl-CoA synthatase or thiokinase.

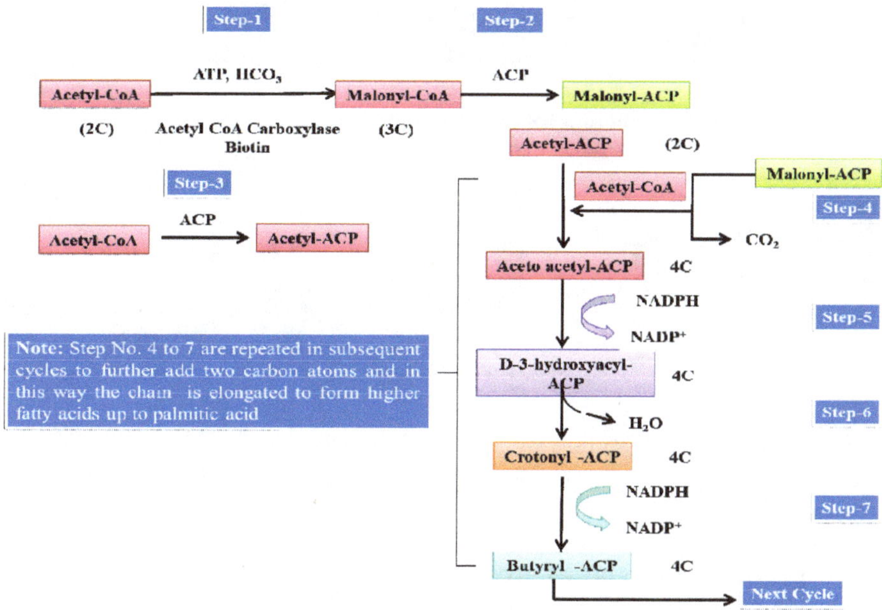

Fig. 6.18: Biosynthesis of fatty acids

However, the process of degradation takes place inside the mitochondria, so the activated fatty acids have to cross the barrier (mitochondrial membrane) to get entry into the mitochondrial matrix as shown in Fig. 6.19. As the mitochondrial inner membrane is impermeable to most acyl-CoA molecules, a special carrier called carnitine plays its role in transporting acyl groups into the mitochondrion. Carnitine acyl derivative is formed by reaction of carnitine acyl transferase-I enzyme with the fatty acyl-CoA. This derivative is capable for crossing the mitochondrial membrane. Inside the mitochondria, fatty acyl-CoA is regenerated by the help of enzyme carnitine acyl transferase-II and the carnitine is transported back into the cytosol by the carrier protein. This cycle is repeated again and again and the activated fatty acids are brought into the mitochondria for oxidation.

Inside mitochondria, the sequence of reactions leading to oxidation of activated fatty acids are as shown in Fig. 6.20.

1. Dehydrogenase.α, β-enoyl-CoA is then hydrolyzed into 3-Hydroxyacyl-CoA by the enzyme enoyl-CoA hydratase.
2. α, β-enoyl-CoA is then hydrolyzed into 3-Hydroxyacyl-CoA by the enzyme enoyl-CoA hydratase.
3. Again, 3-Hydroxyacyl-CoA is oxidized to β-Ketoacyl-CoA by the enzyme L-β Hydroxyacyl-CoA dehydrogenase..

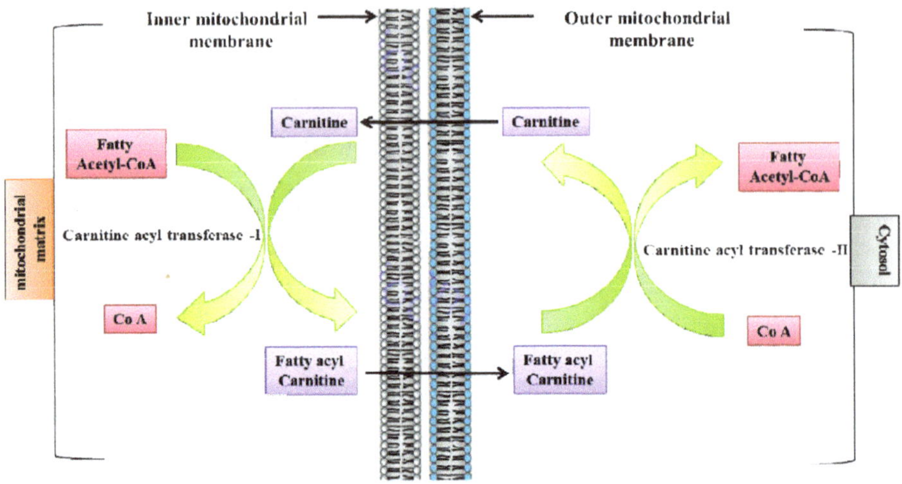

Fig. 6.19: Transport of fatty acids across mitochondrial membrane for oxidation

4. Finally, thiolysis of to β-Ketoacyl-CoA occurs in presence of enzyme thiolaseto yield two molecules- acyl-CoA and Acetyl-CoA.

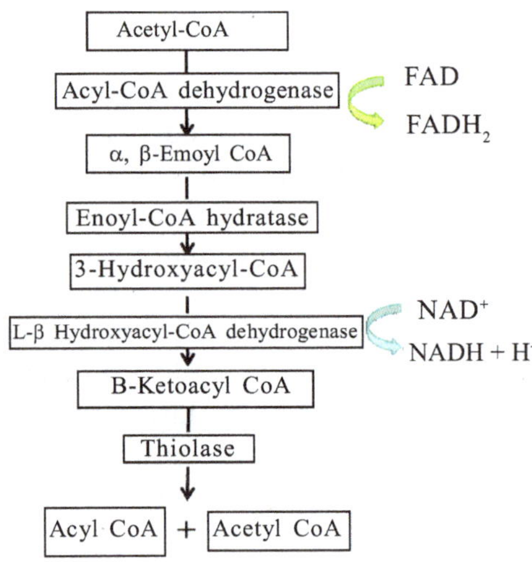

Fig. 6.20: Oxidation of fatty acids

These four steps are repeated again and again to remove two carbon fragments each time.

Ketone bodies

As discussed above, acetyl-CoA is produced as a result of fatty acid oxidation inside mitochondria of liver. This acetyl-CoA is utilized back in fatty acid synthesis or in citric acid cycle. Therefore, a balance in production and utilization of acetyl-CoA is maintained. However, in case of disturbance in this balance, if oxidation of fatty acid is more than its synthesis, then acetyl-CoA is produced beyond the capacity of mitochondria to oxidize it or use it in some other metabolic pathway. This usually happens either when the carbohydrates are not readily available to tissues and fatty acids are oxidized to meet body's energy needs or there is imbalance in the ratio of glucagon (stimulates ketogenesis) and insulin (inhibits ketogenesis) hormones. Therefore, accumulation of acetyl-CoA leads to production of ketone bodies and this process is known as ketogenesis. Three compounds namely, acetoacetate, β-hydroxybutyrate and acetone are referred as ketone bodies.

The synthesis of ketone bodies is as shown in the Fig. 6.21.Thiolase catalyzes the condensation of two molecules of acetyl-CoA to form acetoacetyl-CoA. The latter combines with another molecule of acetyl-CoA to synthesize β-hydroxyl β-methyl glutaryl-CoA (HMG-CoA) in presence of enzyme HMG-CoA synthase. Further, HMG-CoA is cleaved into acetoacetate and acetyl CoA by HMG-CoA lyase. Decarboxylation of acetoacetate leads to production of acetate, whereas its reduction by dehydrogenase produces β-hydroxybutyrate.

Ketone bodies can cross the blood brain barrier and act as a fuel for normal functioning of brain and central nervous system in conditions such as diabetes mellitus and starvation. They also serve as a source of energy for peripheral tissues such as skeletal muscles, heart muscles etc. As these are soluble in water, therefore, their transport from liver to the site of action is quite easy.

The concentration of ketone bodies in a normal human being is usually 0.2mmol/Litre of blood plasma. However, during starvation and diabetes their level increases to 3-5mmol/Litre and 12mmol/Litre, respectively. When ketone bodies production rate is far more than their utilization, then they start accumulating, leading to abnormal state called ketosis. If this condition persists longer, either due to starvation or diabetes (Type-I), then ketone bodies dissociate resulting in lowering of pH. This medical condition is called ketoacidosis and may result in coma and even death. Certain food components like fatty acids and amino acids (leucine, lysine and tyrosine) are ketogenic while glucose, glycerol and glucogenic amino acids are anti ketogenic in nature.

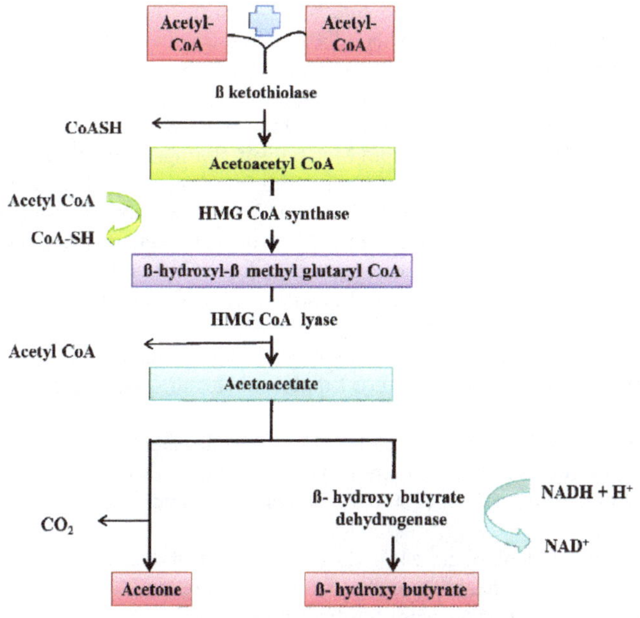

Fig. 6.21: Ketone bodies synthesis

Sample questions

1. Define lipids and classify them.
2. What are phospholipids? Discuss their synthesis and degradation in detail.
3. What are glycolipids? Discuss their metabolism in detail.
4. Classify various lipoproteins on the basis of density.
5. Write a short note on derived lipids.
6. What is cholesterol? Draw its structure.
7. Explain biosynthesis of cholesterol in flow diagram.
8. What are bile acids? Discuss their biochemical function. Write a short note on biosynthesis of bile acids from cholesterols.
9. Write a short note on biosynthesis of fatty acids.
10. Write a short note on fatty acid oxidation.
11. What are ketone bodies? Briefly explain their biosynthesis.
12. Define the terms (1) Prostaglandins (2) Thromboxanes (3) Leukotrienes

Suggested Readings

Carter, R. H. (1981). Lipid Biochemistry: An Introduction, MI Gurr, AT James. Chapman & Hall, London (1980).

Cox, M. M. and Nelson, D. L. (2008). Lehninger Principles of Biochemistry. 5th edn. New York: Worth Publishers.

Eskin, N.M.A. and Shahidi, F. (2013). Biochemistry of Foods. 3rd edn. USA: Academic Press Elsevier.

Gurr, M. I., Harwood, J.L. and Frayn, K.N. (2002). Lipid Biochemistry (Vol. 409). Oxford, UK: Blackwell Science.

Harwood, J. L., Gurr, M. I., Frayn, K. N., Murphy, D. J. and Michell, R. H. (2016). Lipids: Biochemistry, Biotechnology and Health. John Wiley & Sons.

Marinetti, G. V. (2012). Disorders of Lipid Metabolism. Springer Science & Business Media.

Moffatt, R. J. and Stamford, B. (Eds.). (2005). Lipid Metabolism and Health. CRC Press.

Moser, A.B., Steinberg, S.J., Watkins, P.A., Moser, H.W., Ramaswamy, K., Siegmund, K.D., Lee, D.R., Ely, J.J., Ryder, O.A. and Hacia, J.G., 2011. Human and great ape red blood cells differ in plasmalogen levels and composition. Lipids in health and disease, 10(1), p.101.

Satyanarayana, U. and Chakrapani, U. (2013). Biochemistry. 4th ed. Kolkata: Elsevier Books and Allied (P) Ltd.

Singh, N.P. (2005). Biochemistry. New Delhi: Akansha Publishing House.

Stamford, B. A. and Moffatt, R. J. (2005). Lipids and Health: past, present, and future. Lipid Metabolism and Health, 1.

Talwar, G.P., Srivastava, L.M. and Moudgil, K. D. (1989). Textbook of Biochemistry and Human Biology, 2nd edn. New Delhi: Prentice Hall of India Pvt. Ltd.

Vance, J. E. and Vance, D. E. (Eds.). (2008). Biochemistry of Lipids, Lipoproteins and Membranes. Elsevier.

Yan, Y., Borrego, E. and Kolomiets, M. V. (2013). Lipid Metabolism. Baez, RV, Ed, 393-442.

http://www.cholesterol-and-health.com/cholesterol_structure.html

https://themedicalbiochemistrypage.org/bileacids.php

7

Protein Metabolism

Introduction

Proteins are the most abundant organic compounds constituting major part of the body. Chemically, proteins are nitrogen containing macromolecules and polymers of L-α-amino acids, which serve a wide range of structural and biological functions in animal body. For example, connective tissue made up of collagen, metabolic enzymes, neurotransmitters, signaling molecules, structural components of cell membranes and cell organelles, etc. all are chemically proteins.

Before going into the details of protein metabolism, first, we need to understand that there are majorly two sources of proteins in our body: dietary protein and body protein. Dietary protein is taken in diet, while body protein indicates the protein and amino acids synthesized in animal body to fulfill the requirements for growth and development. The digestion of proteins will not be discussed here as it has been described in detail in chapter 12. Dietary protein is digested in presence of several enzymes such as trypsin, pepsin, chymotrypsin, aminopeptidases and carboxypeptidases and is converted into mono-, di- and oligopeptides in stomach and small intestine; and finally converted into amino acids. These amino acids are transported to the system through intestinal mucosa via active transport, from where they enter into the cells for further metabolism. Therefore, 'metabolism of proteins' is more appropriately learnt as 'metabolism of amino acids'.

Metabolic fates and amino acid pool

An adult human body contains approximately 100g free amino acids, which is maintained due to the continual synthesis and breakdown of body protein and other non-protein molecules. Here, amino acid pool does not indicate a particular compartment of amino acids in human body; in fact, these free amino acids are present in every cell for various catabolic and anabolic functions. An overview of metabolic fates of amino acids is shown below (Fig. 7.1):

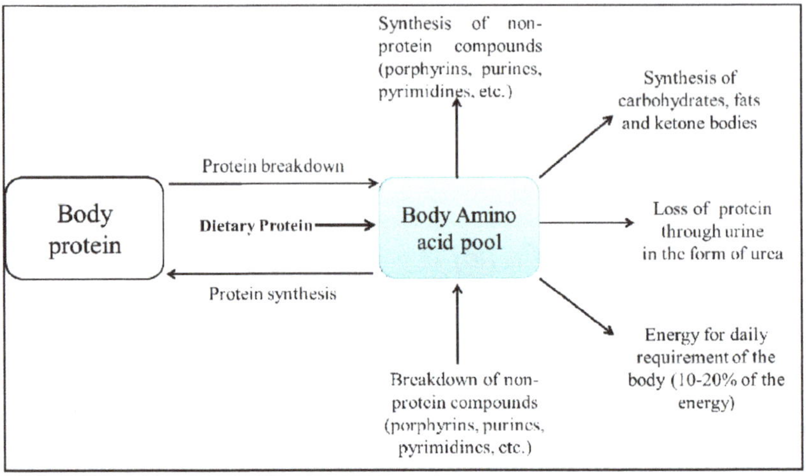

Fig. 7.1: An overview of body amino acid pool and its possible metabolic fates

Amino acid pool and body protein remain in equilibrium due to the continuous breakdown and synthesis of proteins as shown in the Fig. 7.1. Dietary proteins, body protein and breakdown of non-protein components serve as the sources of amino acid pool. The possible fates of amino acids are:

1. Synthesis of body protein for normal growth (structural) and metabolic (biological) functions including enzymes, hormones, immunoproteins, contractile proteins, etc.
2. Synthesis of nitrogen containing non-protein compounds such as purines, pyrimidines, serotonin, etc.
3. In excess, amino acids are converted into carbohydrates, fats and ketone bodies as amino acids cannot be stored like carbohydrates (starch) and fats (triacylglycerols). Amino group is lost as urea to be excreted through urine.
4. Energy production (ATP) for dairy requirement of the body.

Metabolism of amino acids

Metabolism of amino acids involves catabolism (degradation and conversion into other metabolites) and anabolism (new synthesis). There are 20 amino acids out of which ten are essential and need to be taken through diet. Other ten amino acids are non-essential (also known as conditionally essential) and synthesized in body by their own from essential amino acids.In this chapter, we will discuss only the catabolism of amino acids. An overview of the amino acid catabolism is shown in the Fig. 7.2 given below:

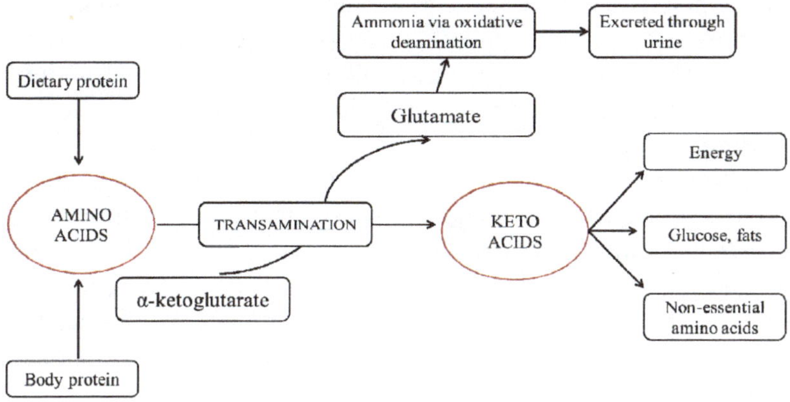

Fig. 7.2: An overview of the amino acid catabolism (metabolism)

All amino acids are catabolized majorly by two reactions: 1) Transamination, followed by 2) Deamination (oxidative and/or non-oxidative).

Transamination

Transamination is the transfer of an amino group (-NH_2) from an amino acid to a keto acid in presence of transaminase (aminotransferase). As a result of transamination, new amino acids and new keto acids are formed. This process is reversible. The enzyme involved in this reaction requires pyridoxal phosphate (PLP) (vitamin B_6) as a co-enzyme. The keto acid formed as a result of transamination can be used for the production of energy through citric acid cycle, or can be converted into glucose and fats, or can be used to synthesize non-essential amino acids (as shown in the Fig. 7.2). An example of transamination reaction is shown below (Fig. 7.3):

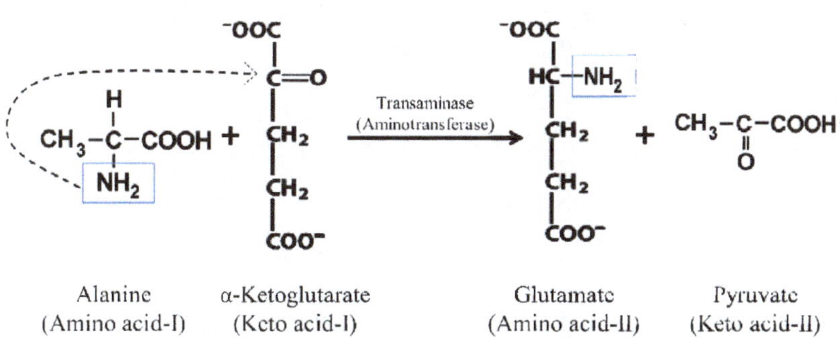

Fig. 7.3: An example of transamination reaction

It should be noted that glutamate and glutamine are the most abundant amino acids (non-essential) in human body, which play a critical role in amino acid metabolism. They serve as a **central collection point** for all amino acids' catabolic reactions. It means that when amino acids (in cytosol of hepatocytes) are transaminated, amino group is finally transferred to α-ketoglutarate (keto acid) resulting the formation of glutamate. Glutamate is the only amino acid which undergoes oxidative deamination. This glutamate enters into mitochondria and releases its amino group to form ammonium ion, which is finally converted into urea and excreted through urine. It should be noted that during transamination no ammonia is generated; it is generated only in further reactions depending on the organ (liver or skeletal muscle) and physiological conditions.All amino acids exceptlysine, threonine, proline andhydroxyproline participate intransamination.

Deamination

Deamination is the removal of amino group from an amino acid to form ammonia. Deamination can be oxidative or non-oxidative.

*1. Oxidative deamination:*Removal of amino group from an amino acid to form free ammonia in presence of L-amino oxidase is known as oxidative deamination.The resultant keto acid can either be used for energy production or a variety of reactions. The reaction is shown below:

$$\text{L-a-amino acid} \xrightarrow{\text{L-amino acid oxidase}} \text{a-keto acid} + NH_3$$

*2. Non-oxidative deamination:*A few amino acids undergo deamination under non-oxidative conditions. For example:

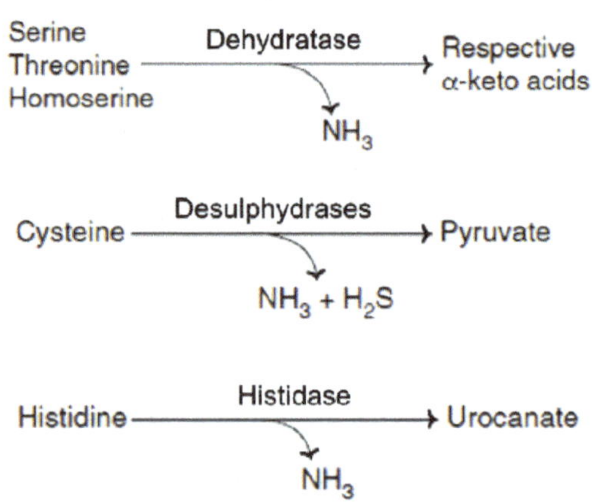

Fig. 7.4: Reactions of non-oxidative deamination of amino acids

As discussed above, after transamination followed by oxidative deamination, most of the amino acids produce keto acids (generally glutamate) and release ammonia. Now, we will discuss the fate of ammonia. In body, ammonia remains in the form of ammonium ions (NH_4^+) at physiological pH (Fig. 7.5). As ammonia is generated in almost all the tissues, it is transported to the liver in the form of glutamine. In other words, it can be said that glutamine serves as a storage form of ammonia; and ammonia is transported from different organs to the liver in the form of glutamine. The conversion of glutamate into glutamine is shown below (Fig. 7.5):

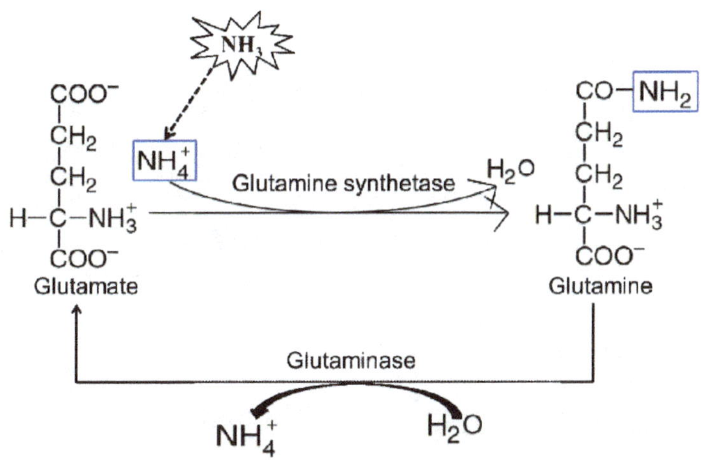

Fig. 7.5: Conversion of glutamate into glutamine and vice-versa.

It should be noted that ammonia generated in muscles from the catabolism of amino acids is transported in the form of alanine and not in the form of glutamine, because in skeletal muscle, excess amino groups are generally transferred to pyruvate to form alanine.

When alanine (from muscles) and glutamine (from other organs) approach towards liver, they are converted into pyruvate and glutamate, respectively, and release ammonia in the form of ammonium ions as shown in Fig. 7.5. Now, it is interesting to note why ammonia is transported in the form of glutamine or alanine from different organs to the liver? Why not in the form of ammonia? The answer is that ammonia is a toxic gas and very harmful for the functioning of the brain. If ammonia is accumulated in body, it may cause slurring of speech, blurring of the vision, coma and even death. Therefore, ammonia is transported in the form of amino acid/s and finally converted into urea (non-toxic) in liver for the excretion through urine. Such organisms in which ammonia is excreted in the form of urea are known as 'Ureotelic', e.g. mammals. Aquatic animals dispose-off ammonia as such and are known as 'Ammoniotelic'. However, in birds and reptiles, ammonia is converted into uric acid for the excretion and are known as'Uricotelic' organisms.

Now, we will see how ammonia is converted into urea? Ammonia is converted into urea through 'Urea cycle', also known as Krebs-Henseleit cycle; which is shown below (Fig. 7.6):

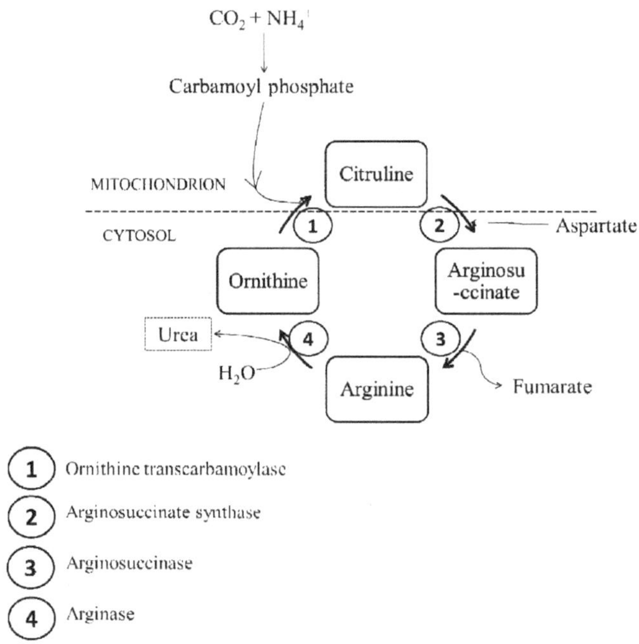

Fig. 7.6: Reactions of Urea cycle

Urea is the end product of amino acid (protein) metabolism. As shown in urea cycle (step 4), arginine is converted into ornithine and urea is released. Thus, urea produced in the liver freely diffuses and is transported to the kidneys via blood; and excreted. A small amount of urea enters the intestine where it is broken down to CO_2 and NH_3 by the bacterial enzyme urease. This ammonia is either lost in the feces or absorbed into the blood.

Sample Questions

1. Give an overview of amino acid pool and its maintenance in human body.
2. Discuss the major reactions of amino acids catabolism with examples.
3. Differentiate between transamination and deamination.
4. What are the possible metabolic fates of amino acids in human body?
5. After deamination, ammonia is transferred in the form of glutamine from different organs to the liver. Why?

6. Give the reactions and importance of urea cycle.
7. Why is ammonia excreted in the form of urea and not in its original form in mammals?
8. Differentiate among uricotelic, ureotelic and ammoniotelic organisms.

Suggested Readings

Bender DA: Amino Acid Metabolism. 2nd ed., John Wiley and Sons, New York, 1985.

Nelson, D.L., Lehninger, A.L. and Cox, M.M., 2008. Lehninger Principles of Biochemistry. Macmillan.

Satyanarayana, U. and Chakrapani, U., 2006. Biochemistry, Books and allied (P) Ltd. Reed Elsevier India Private Limited.

http://www.cuchd.in/e-library/resource_library/University%20Institutes%20of%20Sciences/Fundamentals%20of%20Biochemistry/Chap-26.pdf

https://www.albany.edu/faculty/cs812/bio366/L08_AminoAcidMetab1.pdf

http://www.namrata.co/category/metabolism-proteins/

8

Elementary Knowledge of Milk Synthesis in Mammary Glands

Introduction

The basic physiological structures of mammary glands in all species are almost same and are located externally in order to fulfill the milk requirement of neonates after parturition. The process of secretion of milk from the mammary glands is known as lacto genesis. Milk provides an almost complete nutrition to neonates from birth till weaning period. It is a complex mixture of various constituents like:

1. Lipids such as triacylglycerols (TAG), phospholipids, cholesterols and other sterols, free fatty acids and fat-soluble vitamins
2. Carbohydrates such as lactose, oligosaccharides
3. Proteins such as caseins, α-lactalbumin, β-lactoglobulin, immunoglobulin's, bovine serum albumin, lactoferrin, lysozyme, enzymes, hormones and growth factors
4. Non-protein nitrogenous compounds like urea, creatine, creatinine, uric acid and amino acids
5. Major and minor minerals, water-soluble vitamins and cells (leucocytes, macrophages, and epithelial cells).

Although, these constituents vary between species, breeds, stages of lactation, and even between individuals. Some constituents like vitamins, hormones, and growth factors are quantitatively present in minor amount but has significant biological importance.

The precursors required for milk production are supplied by the bloodstream into the secretory cells of the mammary gland. It is estimated that 1 litre of milk production requires about 400-800 litres of blood to deliver milk precursor into the mammary gland. Some constituents in the milk come unchanged from blood such as minerals, some hormones, immunoglobulins, bovine serum albumin and precursors for milk proteins and carbohydrates. In lactating mammary gland, the primary substrates for milk synthesis are glucose, amino acids, fatty acids, ß-hydroxybutyrate and salt as shown in fig. 8.1.

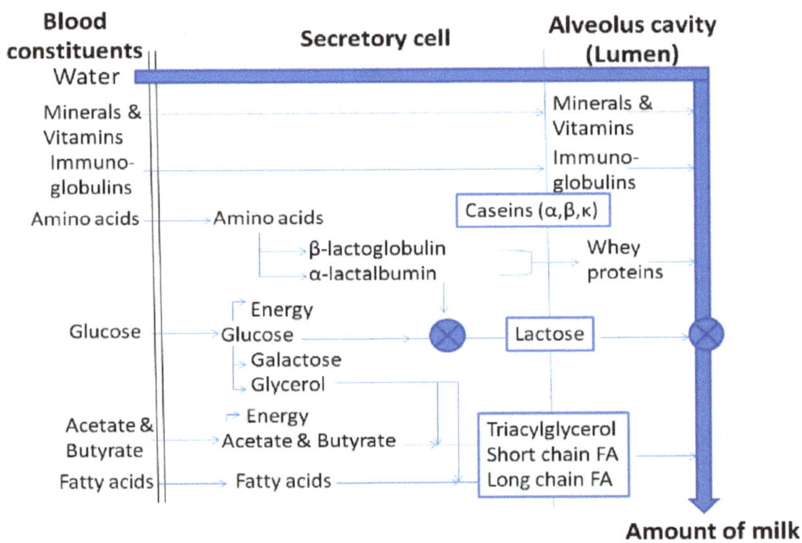

Fig. 8.1: Milk constituents and their source in mammary gland

Milk constituents as such are synthesized from content acquired from the blood in the secretory cells. The secretory cells are present as a layer of spherical or pear shaped bodies known as alveoli, around the lumen. This bunches of alveoli embodies lobules and each lobule is connected and separated by connective tissue for draining constituents into common milk duct. These are conjointly known as 'lobule-alveolar system'. The secretory cells, primarily secrets milk into the alveolar lumen. Lumen is further surrounded by layer of myoepithelial cells. When the lumen is filled completely, every alveolus enclosed by the myoepithelial cells begin to collapse due to release of hormone, oxytocin that enables the milk let down process. Thus, during suckling by newborn or milking of mammary gland, milk is drained through alveolar lumen into milk ducts towards cisterns and reaches teat canal. The cistern precedes the milk to the outside through the teat canal (Fig. 8.2).

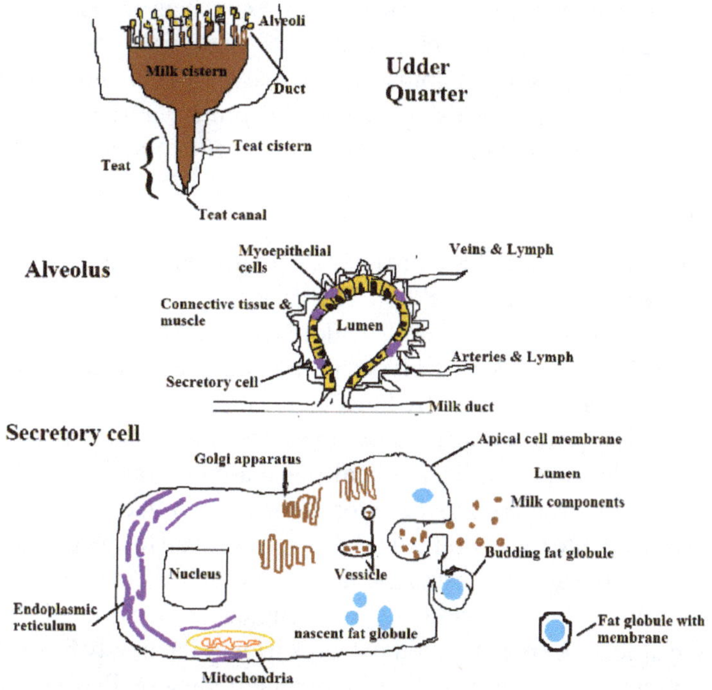

Fig. 8.2: Milk biosynthesis in lactating mammary gland

The milk fractions are synthesized mainly by the endoplasmic reticulum (ER) and attached ribosomes within the epithelial cells and secreted via secretory cells into the lumen. The energy is endowed by the mitochondria to the ER. The constituents are then passed along with the golgi apparatus, from where they head out from the cell in the form of vesicles. The vesicles contain both liquid non-fat components such as casein and liquid fat droplets. These secretory vesicles then goes over the cytoplasm and forces out of apical plasma membrane to be further secreted into the lumen. The apical plasma membrane of secretory epithelial cells eventually emerges as milk fat globule membrane (MFGM).

After parturition, the first form of milk produced by mammary gland in mammals is known as colostrum, which usually last for five days and differs from milk with respect to density, color and acidity. It is mildly acidic (pH 6.4) also. It contains whey protein (immunoglobulin's such as IgA, IgG and IgM) and lymphocytes that safeguards the neonates against diseases. In general, protein substances in colostrum account for approximately 60% of the dry matter, includes 80% or more of whey protein and around 20% casein. Thus, the components are substantially larger than the normal milk.

The milk of different species differs in composition due to the specific nutritional and physiological requirements of their neonates. Thus, great importance is attached to the level of the constituents in milk such as protein, fat, lactose, and minerals, which are likely to change due to a number of factors. Biosynthesis of major constituents such as lipids, lactose and protein in milk requires different sources such as fatty acids, glucose and amino acids from blood. The biosynthesis of these major constituents is further explained in detail.

Biosynthesis of milk lipids

Milk lipid is the most variable constituent, both in concentration and composition. It is majorly influenced by the diet. The triacylglycerol and phospholipids are synthesized primarily on smooth endoplasmic reticulum. They are mainly composed of fatty acids and glycerol and the precursors required for fatty acids synthesis are provided by basolateral membrane of the secretory cells (membrane transport system). These lipids are enveloped with a layer of polar lipids & proteins derived from apical plasma membrane and are released as discrete small fat droplets into the cytoplasm. These numerous small fat globules will coalesce up into the budding lipid droplet and head towards the apical plasma membrane. At the apical plasma membrane it pinches out and enters the lumen as a large lipid droplet. During the lipid synthesis process inside the mammary gland, the apical plasma membrane itself surfaces the newly formed fat globule as a membrane known as milk fat globule membrane (MFGM). The MFGM is known to contain numerous enzymes, including oxidases, reductases, hydrolases and oxidoreductase etc. In the secreted milk, the diameter of milk fat globule ranges from less than 0.1 to more than 22μm which implies that some lipid droplets grow substantially larger and some smaller between the time of their formation and secretion into the lumen. The lipids for normal bovine milk and their composition are represented in Table 8.1.

The biosynthesis of milk lipids (Fig. 8.4) mainly occurs within the epithelial cells through blood or *de novo synthesis* in mammary gland. The glycerol required for triacylglycerol formation comes from blood glucose through glycolysis or by using circulating glycerol which is formed by hydrolysis of triacylglycerol in the body through glyceroneogenesis. The fatty acid ($\geq$ 16 carbons) are incorporated into the milk by *de novo synthesis* and fatty acids ($\leq$ 16 carbons) are preformed and incorporated from diet and body reserve as such in mammary gland. Thus, blood contributes around 40-60% of milk fatty acids, mostly from very low density lipoproteins (VLDL). The lipoproteins basically involved in lipid metabolisms are chylomicrons, low density lipoprotein (LDL) and high density lipoprotein (HDL) and very low density lipoprotein (VLDL) particles. After immediate ingestion of food especially rich in fats; the concentration of chylomicrons in blood increases giving blood serum, a milky appearance. They are released in the blood from thoracic duct of lymph. These all lipoproteins plays important role in transportation of blood lipids to and from different tissues and organs of the body. The blood lipoproteins are of interest from of human health perspective, especially obesity and cardiovascular diseases. Lipoprotein lipase (LPL) found in mammary capillaries breaks up the triacylglycerols into one, two or all three of the fatty acids and results in release of derivatives such as diacylglycerols, monoacylglycerols, free fatty acids and glycerol. The free fatty acids, monacylglycerols, diacylglycerols and glycerol can again be resynthesized into triacylglycerol, a major lipid in epithelial secretory cells.

The blood lipids contributes for 100% of stearic acid (C18:0), linoleic acid (C18:2) and 50% of palmitic acid (C16:0) in ruminant milk. The lipoprotein lipase existent in alveolar blood capillaries hydrolyses blood lipids, its activity enhances eight times on inception of lactation. In ruminant, about 50% of total milk fatty acids are carried from the blood via diet or other organs, accounting around 100% of stearic acid (C18:0), oleic acid (C18:1), linoleic acid (C18:2, an essential fatty acid); and 50% of palmitic acid (C16:0). This palmitic acid (C16:0) or other acids is further lengthened in mitochondrial cell by additions of acetyl CoA or malonyl CoA to the growing chain.

The 30% of principal monounsaturated fatty acid, palmitoleic acid (Cl6:1) and oleic (C18:1) are derived from desaturation of palmitic acid (16:0) and stearic acid (18:0) respectively, due to enzyme activity inside the secretory cells. Some other short chain unsaturated acids like myristoleic acid (C14:1) are probably present due to same enzyme action. The essential fatty acids like linoleic acid (C18:2, n3) and α-linolenic acid (C18:3, n6) cannot be synthesized by mammals and must be obtained through diet. These two essential polyunsaturated fatty acids may then be elongated and further desaturated by mechanisms allied with, stearic acid → oleic acid, to add a range of long polyunsaturated fatty acids. The complete or incomplete δ-oxidation of fatty acids synthesizes δ-Hydroxy acids or β-keto acids.

Table 8.1: Lipids in bovine milk

Classes of lipid	% of the total lipid
Triacylglycerol	95.8
Diacylglycerol	2.25
Monoacylglycerol	0.08
Phospholipid	1.11
Cholesterol	0.46
Free fatty acids	0.28

Source: Jensen & Newburg (1995)

For *de novo* synthesis of fatty acids in mammary gland, a carbon source and reducing equivalents is needed in the form of NADPH + H^+. The principal precursor for biosynthesis of fatty acids is acetyl CoA in milk of all species. It is derived from blood glucose in non-ruminants while in ruminant milk it comes from acetate or β-hydroxybutyrate. The acetate or β-hydroxybutyrate is formed in rumen by micro-organismsand then conveyed to the blood. The mammary gland tissues of ruminant have little ATP citrate lyase activity, which is essential to synthesize lipids from blood glucose. Hence, the blood glucose amount in ruminant milk is preserved for lactose biosynthesis and not used in lipid synthesis. The notable interspecies difference is by virtue of change in precursors of fatty acid and reflected a change in fatty acid profile of milk. The milk lipid biosynthesis is hindered by lack of roughage in the ruminant's diet, which markedly reduces the useable acetate and β-hydroxybutyrate concentration. Ruminant milk displays the large presence of butyric acid due to direct incorporation of butyrate, which is formed from reduction of β-hydroxybutyrate.

In ruminants, inside the cytoplasm, acetyl CoA is first converted to malonyl CoA and this further combined with an 'acyl carrier protein' (ACP) and builds different fatty acids. The thriving fatty acid chain usually esterified to an acyl carrier protein and known by malonyl CoA pathway (Fig. 8.3). Each cycle by

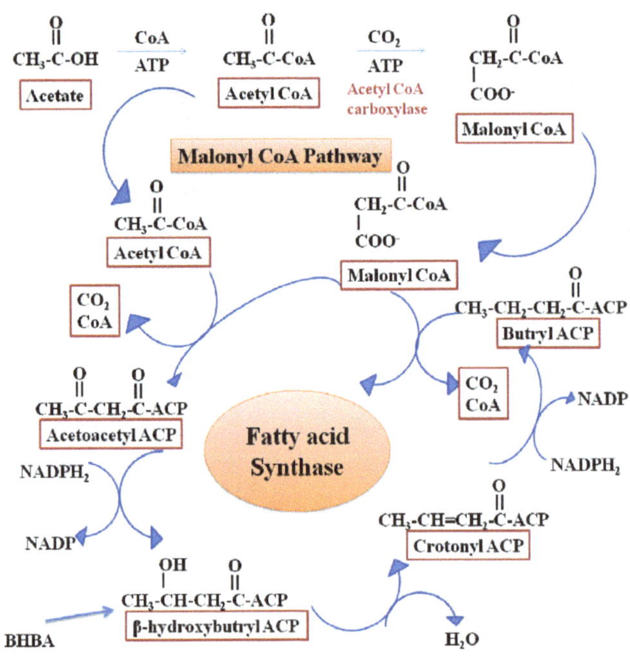

BHBA: β-hydroxybutyric acid, ACP: Acyl Carrier Protein

Fig. 8.3: Fatty acid synthase complex via Malonyl CoA pathway

this pathway results in inclusion of two carbons in fatty acid chain. However, after palmitic acid (C16:0) the further building up of fatty acids through the malonyl CoA pathway not advances. The enzyme thioacylase in the mammary tissue is liable for release of the acyl fatty acid from the carrier protein at any part between C4 and C16. The thioacylase activity may be the reason for plausible inter species disparity in milk fatty acid profiles. About 100% of the capric acid (C10:0), lauric acid (C12:0) and myristic acid (C14:0); and 50% of the palmitic acid (C16:0) are formed through malonyl CoA pathway. While, butyric acid (C4:0), caproic acid (C6:0) and caprylic acid (C8:0) are incorporated in milk from β-hydroxybutyrate or acetate via other pathways.

The enzymes present in endoplasmic reticulum mainly catalyze the reaction for synthesis of the triacylglycerols (TAG) (Fig. 8.3). The fatty acid arranged in a specific combination over a TAG and gives a unique lipid composition; different from the TAG of the other body tissues and plasma or from other common dietary fats and oils. The fatty acids are usually attached at sn-1, sn-2 and sn-3 positions of TAG. The medium chain saturated fatty acids numbered from C12:0 to C16:0 are esterified principally at the sn-2, while short chain saturated fatty acids, C4:0 to C10:0 are esterified at sn-3. However, whenever chain length stretches to C16:0 then esterification incurs at the sn-2; this happens more in human milk fat then bovine. Stearic acid (C18:0) is esterified mainly at sn-1 position. Most of the unsaturated fatty acids are esterified in approximately equal proportion at the sn-1 and sn-3.

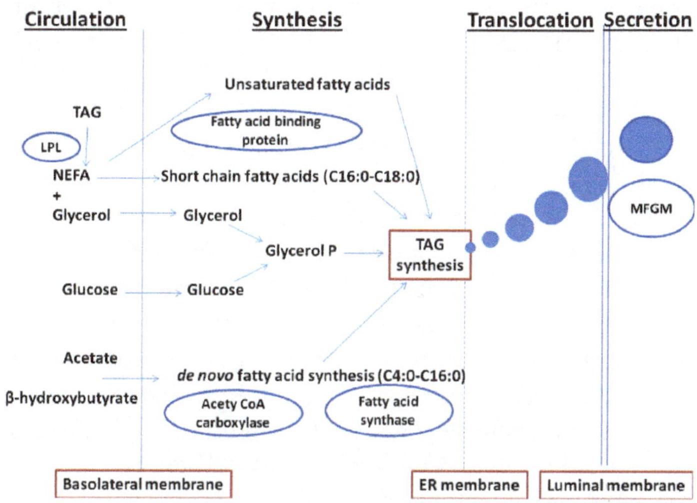

Fig. 8.4: Milk lipid synthesis

NEFA: Non-esterified fatty acids, LPL: Lipoprotein lipase, TAG: Triaylglycerol, MFGM: Milk fat globule membrane

The mechanism for lipid synthesis is such that concentration of each fatty acid especially short chain fatty acid such as C4:0 and long chain unsaturated fatty acid such as C18:1 is so adjusted that they would provide synthesized lipid, a liquid consistency at body temperature due to their low melting point.

Fatty acid distribution over glycerol backbone is of significant importance as it affects textural properties such as melting point or hardness of the milk fat. Thus, several efforts had been employed like transestrification using SnCl or enzymatically to modify texture of fat rich dairy products like butter and anhydrous milk under specified condition.

For effective digestion and assimilation of lipids by human body, the position at which fatty acids are esterified is of significant importance. For example, in human milk, apparently very high proportion of C16:0 is esterified at sn-2 which can be hydrolyzed easily by endogenous lipases and can be metabolized by human infants more efficiently, thus helps in fat absorption. The fatty acids at sn-1 and sn-3 are hydrolyzed by pancreatic lipase, thus, releases short chain saturated fatty acids from C4:0 to C8:0 rapidly in body. These are water soluble fatty acids, hence can be easily absorbed from intestine. The medium and long-chain fatty acids are absorbed as monoacylglycerols rather than as free fatty acids; this turns upto be utterly important for newborns that initially have no bile salts and immature digestive system.

Biosynthesis of milk proteins

The protein synthesis starts as administered from DNA to mRNA (transcription) and mRNA to protein (translation) in all organisms. Mammary specific genes turn on when milk synthesis is active. A strand of messenger RNA (mRNA), separates from DNA inside the nucleus and carries the code for formation of specific protein. As this code comprises of three nucleotides hence named as a triplet codon, which is recognized by anticodon of transfer RNA (tRNA). The specific proteins are formed when each codon in the mRNA comes in position at ribosomes and tRNA transfers amino acid to it. The tRNAs are specific for each amino acid. Thus, the milk protein biosynthesis requires amino acids and energy. Through specific amino acid transport system, amino acids absorbed via basolateral membrane. Inside the secretory cell, ribosomes attached to rough endoplasmic reticulum (RER) restrained the mRNA and bind the available amino acid covalently with the help of (tRNA) to form proteins. These amino acids are activated by reaction with ATP in cytoplasm. Then, they are transported from RER to Golgi apparatus for post-translational processing and packaging as secretory vesicle. These secretory vesicles of milk proteins (with lactose also) move to apical plasma membrane and fuses with inner surface of apical plasma membrane, secretory vesicles then open up and its contents are discharged into alveolar lumen. The protein synthesis required high energy and after ion transport, it is one of the most strenuously expensive processes in the cell. The amount and composition of milk proteins is mostly determined by the genetics of the animal, and is rarely changed through diet. However, immense energy requirement during biosynthesis of milk protein affects the milk protein yield. A variety of milk proteins are synthesized from amino acids inside the mammary gland. The milk proteins synthesized include casein and whey proteins (β-lactoglobulin, α-lactalbumin), membrane bound proteins or enzymes. Casein is the primarily important milk protein, both quantitatively and nutritionally. In milk, it is present as cluster of spherical sub micelles known as casein micelles. Calcium bound to milk casein originates from the plasma in the cytoplasm present on the basolateral membrane. Then, from the cytoplasm taken into secretory vesicles by an ATP-dependent Ca^{2+} pump in golgi apparatus and secretory epithelial membrane. Inside the secretory vesicles, casein and calcium collectively forms large micellar structures as well as complexes with phosphate and citrate to efficaciously reduce Ca^{2+} levels. The phosphates are partially obtained from the hydrolysis of UDP-galactose during the lactose synthesis. Citrate is formed endogenously within the cytoplasm in the cell of alveoli and conveyed into the golgi lumen. Thus, casein micelles contain calcium and inorganic phosphate and small amounts of citrate and magnesium, thereby supplying essential constituents for skeletal development. Immunoglobulins (Igs), is another important milk protein and is found particularly in high concentrations in colostrum. These are selectively transferred from blood or synthesized locally by cells of the immune system. Immunoglobulins bind to specific receptors on basolateral membrane into vesicles and then transported by fusing with inner apical plasma membrane and releases into alveolar lumen. The young mammals depend on the passive transfer of immunoglobulins (antibodies) from their mothers to protect against infection or disease, until their immune system become mature. The shift of immunoglobulins may also incur in-utero across the placenta. Serum albumins do not have receptor but passes along with immunoglobulin, plus they too provide protection from diseases. Milk contains other components that can also work similarly such as lactoferrin, lysozyme, and lactoperoxidase.

Biosynthesis of milk carbohydrates

Lactose is the major carbohydrate found in almost all mammalian milk. It is a disaccharide, synthesized inside golgi apparatus from UDP galactose and glucose, similar to milk protein synthesis. Glucose required is absorbed through blood for lactose synthesis. In a four enzyme Leloir pathway, one molecule of glucose is linked with UDP-galactose, in a reaction catalyzed by the enzyme, lactose synthase that catalyzes the end step in lactose synthesis. This enzyme consists of two components i.e., A and B. The component A is enzyme, galactosyl transferase, non-specific in nature and occurs in several tissues, and conveys the galactose from UDPGal to a number of acceptors. The component B is α-lactalbumin, a protein unique to the mammary gland. It has no enzyme activity but functions to increase the affinity of galactosyl transferase for glucose and also known as enzyme modifier. In the presence of α-lactalbumin, the galactosyl transferase becomes largely specific for glucose, allowing the formation of lactose. The concentration of lactose in milk of various species is directly proportional to the concentration of α-lactalbumin. Due to this reason, some marine mammal's milk neither have α-lactalbumin nor lactose. The high amount of lactose is found in the golgi apparatus during lactation, water is osmotically drawn into it that maintains the fluidity of milk. Glucose emerges into the cell of the basolateral membrane through blood by a specific transport mechanism. Some of the glucose is tranformed to galactose. Glucose and galactose both move in the golgi apparatus and formation of lactose takes place. The lactose and water is secreted through the secretory vesicles accompanied with the milk proteins into the alveolar lumen (Fig. 8.5).

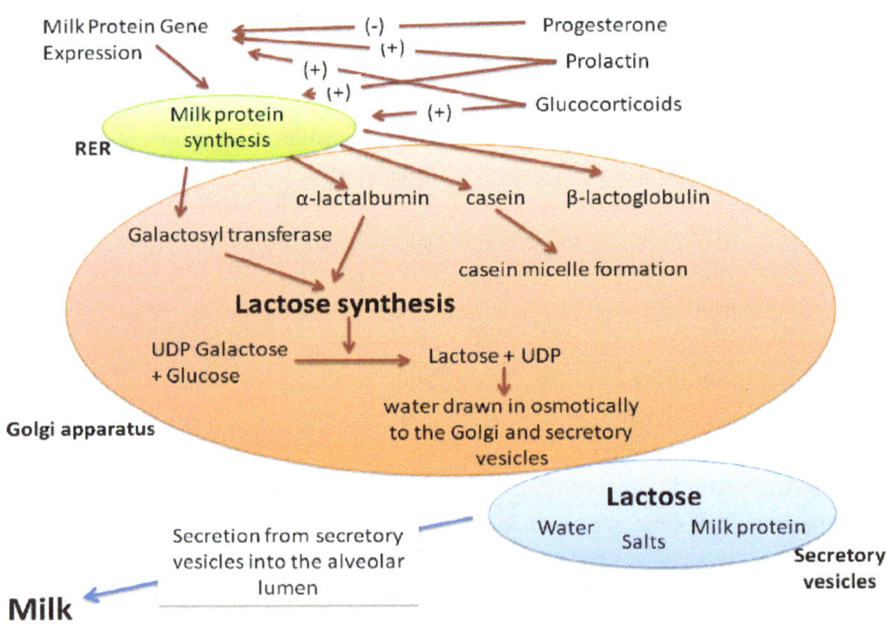

Fig. 8.5: Biosynthesis of lactose and protein

Lactose and NaCl are major decisive fractions of the osmotic pressure in milk inside mammary gland and remains isotonic with blood. During mastitis infection or in late lactation milk, concentration lactose is reduced significantly and NaCl concentration increases. Thus, the osmotic pressure can be regulated during influx of NaCl in mammary gland. The ability to manage osmotic pressure is important enough to uphold an elaborate control mechanism. This control mechanism helps both ruminants and non-ruminants to stop the lactose biosynthesis whenever required.

Secretion of milk from mammary gland

Milk proteins are synthesized on the RER from where they are integrated into the golgi apparatus. Other solid-non-fat constituents such as lactose and salts are also synthesized inside golgi apparatus, similar to protein. The secretory vesicles from the golgi apparatus, separates out and head towards the apical plasma membrane where the membrane surfacing around each vesicle merges with the apical plasma membrane and its constituents are disposed into the alveolar lumen.

Unlike, milk protein and lactose, milk lipids follow a distinct secretion pathway. Lipid synthesis comprises of *de novo* synthesis and is acquired from blood lipids as such. Several small lipid droplets formed enlarges as they head from the endoplasmic reticulum and approaches the apical membrane where they forces out as milk fat globules enrobed in an envelope made of apical plasma membrane which is known as milk fat globule membrane.

These individual constituents are kept serrate inside secretory cell and are not let out until these constituents reached the alveolar lumen where they are mingled as milk.

Sample Questions

1. What are different constituents of milk and their sources for synthesis in mammary gland?
2. What is lobule alveolar system?
3. Briefly explain the process for biosynthesis of milk in mammary gland.
4. How milk lipids are synthesized in mammary gland?
5. Give a brief account of *De novo* synthesis of milk lipids?
6. What milk synthesis primers for major milk constituents?
7. What are the principal precursors for lipid synthesis in ruminant and non-ruminants?
8. Briefly explain the source milk fat globule membrane (MFGM) in fat globule during lipid synthesis?
9. How consistency or liquidity of milk is maintained through fatty acids?

10. Give a brief account of different fatty acids and their esterification position on triacylglycerols during lipid biosynthesis.
11. Explain in detail the biosynthesis of lactose in mammary gland?
12. What is the role of α-lactalbumin in lactose biosynthesis?
13. Why α-lactalbumin is called as enzyme modifier?
14. How lactose affects the osmolarity of lactating mammary gland?
15. What is source of minerals in lactating mammary gland?
16. How milk proteins are synthesized in mammary gland?

Suggested Readings

Aschaffenburg, R. (1961) 'Inherited casein varients in cow's milk'. *Nature (Lond.)* 176, pp. 218.

Eskin, N. A. M. and Goff, H. D. (2013) 'Milk', in Eskin, N. M. A.and Shahidi, F. (2013). *Biochemistry of foods*. 3rd edn. USA: Academic Press Elsevier, pp. 187-208.

Fox, P. F. and McSweeney, P. L. H. (1998). *Dairy Chemistry and Biochemistry*. New York: Blackie Academic and professional (Chapman and Hall).

Fox, P. F. and McSweeney, P. L. H. (2003). *Advanced Dairy Chemistry-1. Proteins*. 3rd edn. New York: Kluwer Academic.

Fox, P. F. and McSweeney, P. L. H. eds. (2006). *Advanced Dairy Chemistry-2. Lipids*. 3rd edn. New York: Kluwer Academic.

Fox, P. F. and McSweeney, P. L. H. eds. (2009). *Advanced Dairy Chemistry-3. Lactose, Water*

9

Human Nutrition

"To eat is a necessity, but to eat intelligently is an art."

-La Rochefoucald

Introduction

Nutrition, a basic human need, is a critical part of health and development. Nutrition can be defined as the intake of food, deliberated in relation to the body's dietary needs. On one hand, good nutrition – an adequate, well balanced diet combined with regular physical activity – is a foundation of good health, whereas, poor nutrition leads to reduced productivity, impaired mental and physical development, increased susceptibility to disease, and reducedimmunity. According to WHO, under-nutrition, together with vitamin and mineral deficiencies accounts for one-third of all child deaths and impairs healthy development and lifelong productivity. At the same time, sedentary lifestyles, demographic shifts, urbanization and the liberalization of markets have steered the world population to being overweight along with risk of chronic diseases like hypertension, dyslipidemia, type 2 diabetes, coronary heart disease, stroke, etc.; thus, resulting in a double burden of malnutrition.

To address this dual burden of malnutrition, individuals need access to sufficient and good quality food; but focussing solely on food security is unlikely to solve the problem of global malnutrition. The challenges of malnutrition cannot be solved just by increasing food production in order to improve health and nutritional status of the population. To resolve such challenges, awareness about food choices for optimal health plays a pivotal role among population.

Human nutrition is the science of understanding food; its component; intake; and how it is related to health, growth, development and diseases. Human nutrition involves the study of ingestion, digestion, absorption and assimilation of the nutrients; their metabolic functions, deficiency or excess disease as well as their regulation through several factors such as choosing particular food components/nutraceuticals or physical activity.

Nutrients: The basics

People eat food, not nutrients; however, it is the combination and amount of the nutrients in the foods consumed that determines our health. Hence, to understand nutrition one must know the basics and terminology used in this field.

- **Food:**Any nutritious substance that people or animals eat or drink; or that plants absorb in order to maintain life and growth.
- **Nutrient:**A component imparting nourishment in order to fulfil the energy needs for growth and development of the body.
- **Nutrition:**Nutrition is the intake of food to fulfil the dietary requirements of the body.
- **Malnutrition:**Malnutrition indicates the deficiency, over-consumption or imbalance of the dietary nutrients leading to health problems. It includes two major terms:
    - Undernutrition: It shows the dietary deficiency of macro- and micronutrients due to lower consumption of food, poor food choices or lesser absorption of nutrients leading to stunted (low height for age), wasting (low weight for height) and underweight (low weight for age) conditions.
    - Overweight/Overnutrition: It is caused by the over consumption of nutrients particularly macronutrients and lesser physical activity than required; leading to obesity and diet-related non-communicable diseases.
- **RDA:**The Recommended Dietary Allowances (RDAs) are quantities of nutrients in the diet that are required to maintain good health in people. In India, RDAs are established by the Indian Council of Medical Research (ICMR), and is revised every few years. A separate RDA value exists for each nutrient and refers to the amount of nutrient expected to maintain good health in people classified based on work, weight, sex and age.

Nutrients cannot be studied separately (in isolated form) as they interact with each other and several other components in the body. Nutrients, after ingestion, passes through gastro-intestinal tract, absorbed via mucosal cells, transported through blood and finally enter into cells and cell organelles for oxidation and other metabolic functions. During this whole journey, nutrients interact with several enzyme molecules, signalling molecules, carrier proteins, lipoproteins and other nutrient molecules. Therefore, the nutrients should be studied by an integrated approach from the perspective of total body functions.

There are more than 50 known nutrients in food which are believed to have a positive impact on human health and development. On the basis of their contribution to the body weight and energy, nutrients are broadly classified into macronutrients and micronutrients. Macronutrients involve nutrients which are generally taken on larger amount and which provide energy through oxidation and helps in growth of the body, e.g. carbohydrates, fats and proteins. Micronutrients consist of vitamins and minerals, taken in minute amount in diet and essential for proper functioning of macronutrients.

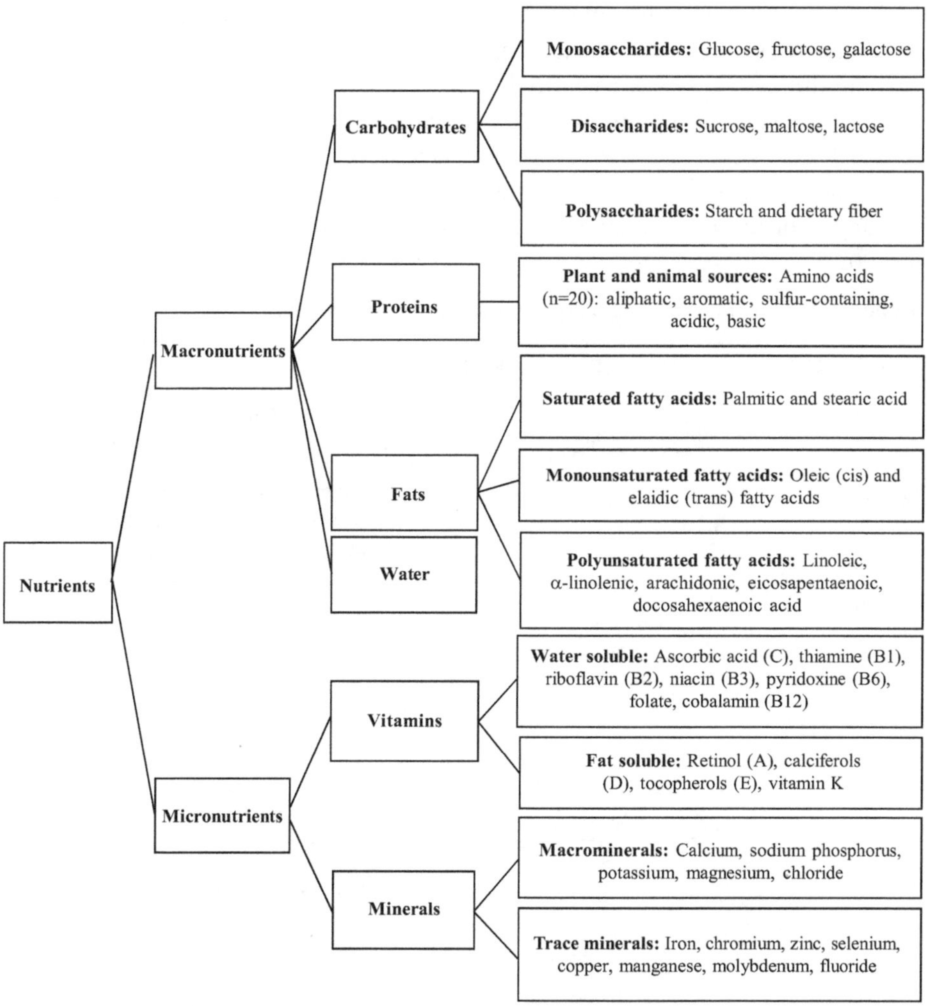

Fig. 9.1: Classification of nutrients

Significance of macro-nutrients in body

Macro-nutrients are needed in larger quantities (in grams) and include carbohydrates, fat, protein and water. Except water the others are also called energy-providing nutrients which are essential for the growth, repair, development and regulation of the life process.

1. Carbohydrates

Carbohydrates, one of the four major macro-nutrients, are polyhydroxy aldehydes or ketones, and are constructed from the atoms of carbon, oxygen, and hydrogen. The proportion of these atoms comes close to that of a "hydrate of carbon" CH_2O, and are thus, termed as'carbohydrates'. Carbohydrates supply 4 kilo calories of energy per gram and consist of two major classes: simple carbohydrates (monosaccharides like fructose, glucose and galactose; disaccharides like sucrose, maltose, lactose and trehalose) and complex carbohydrates. Complex carbohydrates include oligosaccharides (Raffinose, verbascose, stachyose, inulin, fructo and galacto-oligosaccahrides), containing 3 to 10 saccharide units and the polysaccharides containing more than 10 monosaccharide units. Table 9.1 presents a brief overview of the most common dietary sources for the different types of carbohydrates along with their structures.

Table 9.1 Different types of carbohydrates and their dietary sources

CARBOHYDRATES AND THEIR SOURCES		
Fructose	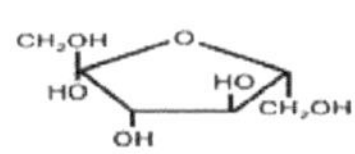	Fruits, vegetables and honey and is derived from the digestion of sucrose
Glucose	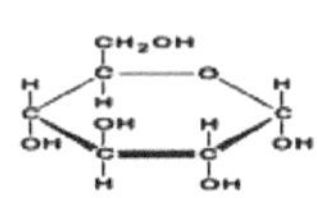	Small amounts are found in fruits, vegetables and honey, manufactured foods and digestion and conversion of other carbohydrates
Galactose	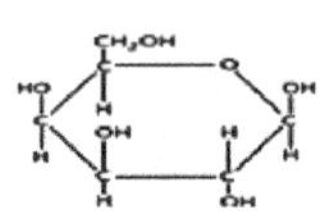	Digestion of lactose
DOSACCHARIDE		
Sucrose	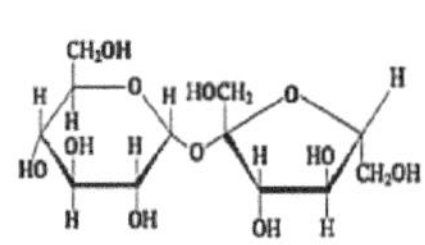	Derived from sugar cane, sugar beet, beetroot and carrots, table sugar and manufactured foods
Maltose	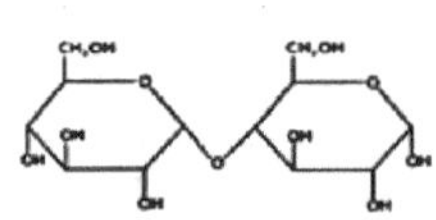	Malted wheat and barley, malt extract and beer
Lactose	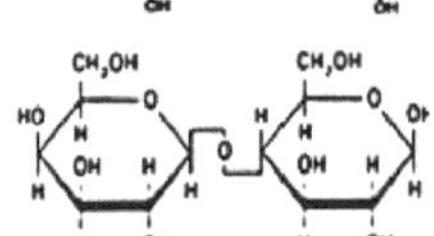	Milk and Milk products
Trehalose	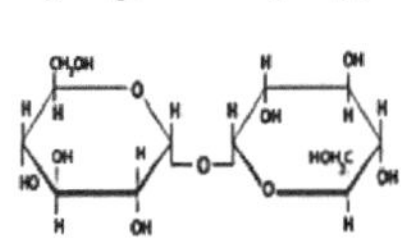	Mushrooms and edible fungi

OLIGOSACCHARIDES	
Raffinose, stachyose, verbascose, inulin, fructo and galacto-oligosaccharides	Legumes, onion, artichoke, fennel, asparagus and Pre-biotics
Starch 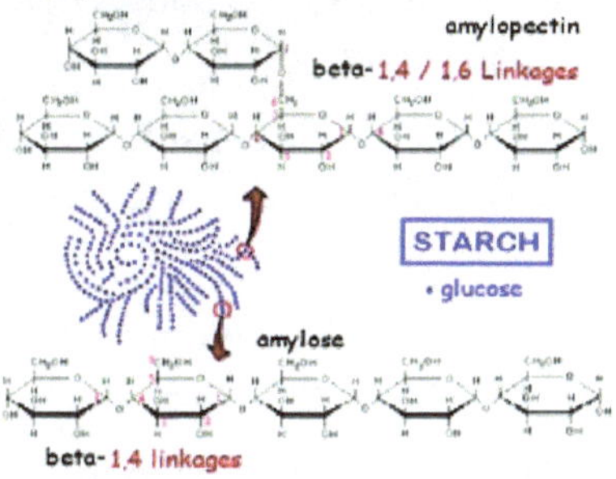	Cereals, potato, and small amounts in other root vegetables and unripe fruits
Non-starch polysaccharides	Vegetables, fruit, whole grain cereals and pulses

Major functions of carbohydrates are summarized in Fig. 9.2 and discussed below:

1. **Major sources of energy:** Carbohydrates are the principal source of energy and fulfils approximately 60–80% of the caloric requirements of the body.
2. **Protein sparing action:** Proteins have to accomplish specific function of body building and growth and hence, carbohydrates spare the proteins from being misused for caloricpurpose.
3. **Absolute requirement of brain:** The brain and other parts of central nervous system are dependent only on glucose for energy, wherein deficiency of glucose for a long time may lead to irreversible brain damage.

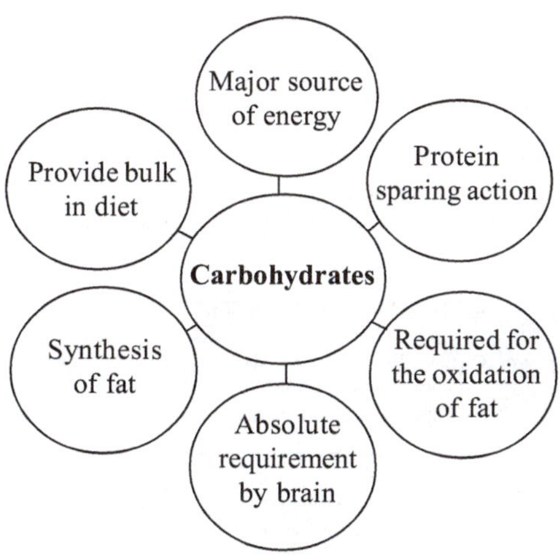

Fig. 9.2: Functions of carbohydrates

4. **Required for the oxidation of fat:** For further oxidation of acetylCoA (product of fatty acidoxidation) via citric acid cycle, acetyl CoA combines with oxaloacetate (derived from carbohydrates) and consequently it can be stated that 'Fatburns in a fuel of carbohydrate'.

5. **Synthesis of pentoses:** Pentoses (e.g. Ribose), the constituents of nucleic acids (DNA, RNA), coenzymes (NAD+, FAD) are produced in carbohydrate metabolism.

6. **Synthesis of fat:** Excess consumption of carbohydrates leads to fat formation and is stored in the adipose tissue.

7. **Importance of non-digestible carbohydrates:** Owing to the absence of enzymes necessary to digest cellulose, hemicellulose, lignins, oligosaccharides, pectins, gums, waxes and other polysaccharides in human body, they remain undigested and contribute to faeces. As they are indigestible they are termed as dietary fibre, nevertheless, as a source of fiber it assumes significance in human body as a bulking agent and a potential energy source for some intestinal bacteria that are able to digest it. Other important functions of dietary fiber are summarized in Figure 9.3.

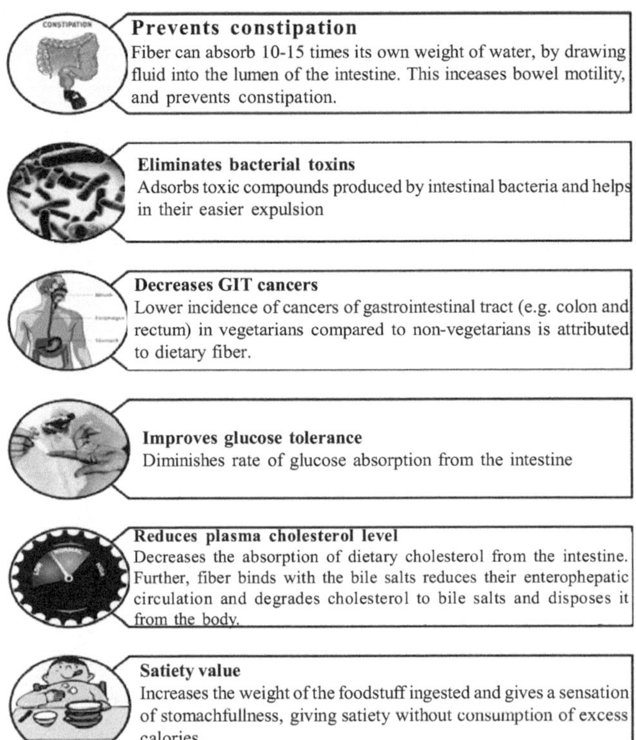

Fig. 9.3: Functions of dietary fiber

2. Proteins

Proteins hold a significant place in nutrition and health and it's evident from the presence of proteins in every cell. Proteins constitute nearly 50% of the dry cell weight. The term protein is derived from a Greek word *proteos* meaning "primary" or "taking first place" which highlights the importance of this nutrient in human nutrition. Chemically, proteins are organic molecules made up of amino acids. Proteins play a variety of functions in body including growth; metabolic reactions; as enzymes, signalling molecules and carriers of nutrients and other metabolites, etc.

Amino acids

Amino acids are the building blocks of proteins containing two functional groups—amino (—NH_2, basic) and carboxyl group (—COOH, acidic). For the synthesis of proteins and other biological functions, the body needs 20 amino acids. Nevertheless, all of them need not be supplied in the diet, hence, nutritionally or based on the essentiality of amino acids they are classified into two groups; (a) essential amino acids and (b) non-essential amino acids (Figure 9.4).

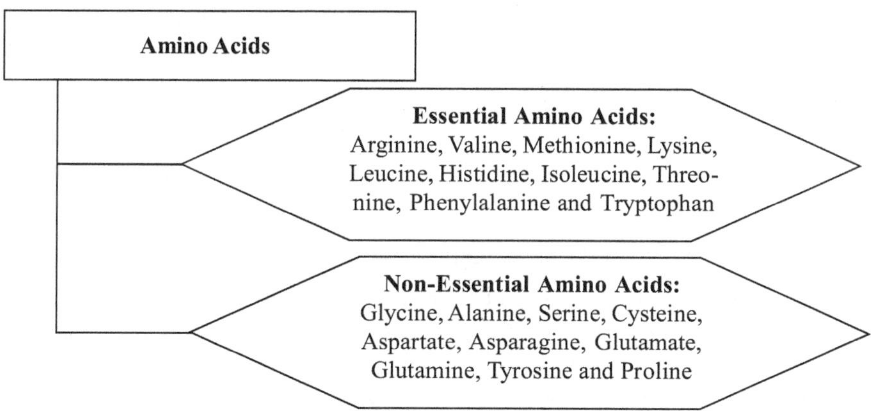

Fig. 9.4: Nutritional classification of amino acids

Functions of proteins

Proteins form a major part of total body structures. They are significantly involved in the formation of cell and plasma membrane, connective tissues, muscles and help in maintaining the integrity of the cell organelles and body parts. It is interesting to note that more than 40% of total body protein is found in skeletal muscle and over 25% found in body organs. Important functions of proteins/amino acids are presented in Table 9.2.

Table 9.2: Functions of proteins in human body

Role	Example	Function
Digestive enzyme	Amylase, lipase, pepsin	Breaks down nutrients in food into small pieces that can be readily absorbed
Transport	Hemoglobin	Carries substances throughout the body in blood or lymph
Structure	Actin, tubulin, keratin	Builds different structures, like the cytoskeleton
Hormone signalling	Insulin, glucagon	Coordinates the activity of different body systems
Defence	Antibodies	Protects the body from foreign pathogens
Contraction	Myosin	Carries out muscle contraction
Storage	Legume storage proteins, egg white (albumin)	Provides food for the early development of the embryo or the seedling
Fluid & electrolytebalance	Albumin	Helps to maintain the volume and composition of body fluids
Acid-base balance	Hemoglobin	Helps maintain the acid-base balance of the body fluids by acting as buffers
Energy	—	Provides fuel for the body's energy needs [4 Kcal/g]
Contribute to sensory & physical properties of food	—	Imparts colour, flavour, odour and texture to foods

3. Lipids

Lipid, a complex molecule, is composed of a mixture of fatty acids and glycerol. Lipids are the concentrated source of energy and integral part of the diet. Just like carbohydrates, it contains carbon, hydrogen and oxygen but contains more carbon and hydrogen andless oxygen. Fats are the important source of essential fatty acids and fat soluble vitamins, pigments and several bioactive components such as such as γ-oryzanol, carotenoids, tocotrienols, etc.

Functions of lipids

a) Lipids are the concentrated source of energy and act as storage form also.

b) Important components of plasma membrane of cells and cell organelles.

c) Only source of fat soluble vitamins such as vitamin A (retinol), vitamin E (tocopherols) and vitamin K.

d) They act as precursors of steroidal hormones and prostaglandins which act as regulators for several metabolic reactions.

e) Lipids act as insulating materials and protects internal organs, in addition to providing shape and smooth appearance to the body.

f) Lipids being present in cell membranes regulate cell permeability and act as carrier and signalling molecules also.

g) Cholesterol, which is a non-saponifiable component of fat (of animal origin), acts as the precursor of steroidal compounds, vitamin D and bile acids. It also helps in maintaining plasticity of plasma membrane.

h) Lipoproteins, complexes of lipids and proteins, play a significant role in absorption and transportation of lipid molecules in body.

i) Lipids are the only source of essential fatty acids such as linoleic and linolenic acid.

4. Water

Water is the most abundant compound in human body and is an essential nutrient as it is required in amounts that exceed the body's ability to produce it. Depending on lean body mass, water accounts for approximately 50-80% of body weight. Water is required for digestion, absorption, transportation, dissolving nutrients, elimination of waste products and thermoregulation.

Functions of water

1. Water is present in each cell, tissues and compartments and acts as a building material.
2. Water serves as a universal solvent for nutrients and helps in transportation of nutrients to cells and in removal of waste products from the cells.
3. It helps in maintaining the fluid and electrolyte balance in body.
4. It helps in maintaining ionic balance in the body.
5. Water maintains the structure of large molecules e.g. proteins and glycogen.
6. It participates in all biochemical reactions.
7. Water limits the changes in body temperature in a warm or cold environmentand thus contributes to thermoregulation.

Significance of micronutrients in body

Unlike macronutrients, micronutrients (vitamins and minerals) do not provide energy and required in very minute amounts. Their foremost function is to assist in chemical reactions of the body; and hence, are extremely vital for the normal functioning of the body.

1. Vitamins

Vitamins are essential for normal metabolism, growth and development, and regulation of cell function and thus, named as vitamins. Vitamins are categorised into fat-soluble (A, D, E and K) and water-soluble (B and C). Kindly refer Chapter 3 for details.

2. Minerals

Minerals are found in ionized form and act mostly as co-factors, which are essential for the functioning of enzymes in the body. Based on their requirement, they are further classified into macrominerals and trace minerals. Macrominerals include calcium, sodium, phosphorus, potassium, magnesium, chloride whereas, traceminerals include iron, chromium, zinc, selenium, copper, manganese, molybdenum, fluoride. A summary of the major characteristics of macrominerals and trace minerals is given in Tables 9.3 and 9.4, respectively.

Table 9.3 Major characteristics of macrominerals

Minerals	Physiological functions	Deficiency diseases/symptoms	Food sources
Calcium	Structural component of bones and teeth; role in muscle contraction, cellular processes, enzyme activation and blood clotting	Rickets, osteomalacia, osteoporosis, tetany	Milk, milk products, sardines, clams, oysters, turnip and mustard greens, broccoli, legumes, dried fruits
Sodium	Water pH and electrolyteregulation; nerve transmission, muscle contraction	Anorexia, nausea, muscle atrophy, poor growth, weight loss	Table salt, meat, seafood, cheese, milk, bread, vegetables (abundant in most foods except fruits)
Phosphorus	Structural componentof bone, teeth, cellmembranes, phospholipids, nucleic acids, nucleotide coenzymes, ATP-ADP phosphate transferring system in cells, pH regulation	Neuromuscular, skeletal, hematologic, and cardiac manifestations; rickets, osteomalacia	Meat, poultry, fish, eggs, milk, milk products, nuts, legumes, grains, cereals
Potassium	Water, electrolyte, and pHbalances; cell membrane transfer	Muscular weakness, mental apathy, cardiac arrhythmias, paralysis, bone fragility	Avocado, banana, driedfruits, orange, peach, potatoes, dried beans, tomato, wheat bran, dairy products, eggs
Magnesium	Component of bones; role in nerve impulse transmission, protein synthesis; enzyme cofactor	Neuromuscular hyper excitability, muscle weakness, tetany	Nuts, legumes, wholegrain cereals, leafy green vegetables
Chloride	Primary anion; maintains pH balance, enzyme activation, component of gastric hydrochloric acid	In infants: loss of appetite, failure to thrive, weakness, lethargy, severe hypokalemia, metabolic acidosis	Table salt, seafood, milk, meat, eggs
Sulphur	Component of sulphur containing amino acids, lipoic acid, and 2 vitamins (thiamin, biotin)	Unknown	Protein foods—meat, poultry, fish, eggs, milk, cheese, legumes and nuts

Table 9.4: Major characteristics of trace minerals

Minerals	Physiological functions	Deficiency diseases/symptoms	Food sources
Iron	Component of haemoglobin andmyoglobin for O_2 transport andcellular use	Listlessness, fatigue, anaemia, palpitations, sore tongue, angularstomatitis, dysphagia, decrease dresistance to infection	Organ meats (liver), meat, molasses, clams, oysters, nuts, legumes, seeds, green leafy vegetables, dried fruits, enriched/whole-grain breads/cereals
Chromium	Normal use of blood glucose andfunction of insulin	Glucose intolerance, glucose and lipid metabolism abnormalities	Mushrooms, prunes, asparagus, organ meats, whole-grain breads and cereals
Zinc	Energy metabolism, metabolism, protein synthesis, collagen formation, alcohol detoxification, carbon dioxide elimination, sexual maturation, taste and smell functions	Poor wound healing, subnormalgrowth, anorexia, abnormal taste/smell; changes in hair, skin,nails; retarded reproductive system development	Oysters, wheat germ, beef, liver, poultry, whole grains
Selenium	Protects cells against destruction by hydrogenperoxide and free radicals	Myalgia, cardiac myopathy,↑ cell fragility, pancreatic degeneration	Grains, meat, poultry, fish, dairyproducts
Copper	Utilization of iron stores, lipids, collagen, pigment, neurotransmitter synthesis	Anaemia, neutropenia, bone abnormalities	Liver, shellfish, wholegrains, legumes, eggs, meat, fish
Manganese	Brain function, collagen, bone,growth, urea, synthesis, glucose and lipid metabolism, CNS function	In animals, possibly humans:impaired growth, skeletal abnormalities, impaired CNS function	Wheat bran, legumes, nuts, lettuce, beetroot, blueberries, pineapple, seafood, poultry, meat
Molybdenum	Metabolism of purines, pyrimidines, pteridines, aldehydes, and oxidation	Hypermethioninemia, ↑ urinaryxanthine, sulfite excretion, ↓ urinarysulfate and urate excretion	Soybeans, lentils, buckwheat, oats, rice, bread
Fluoride	Maintenance of teeth and bone structure	Dental caries, bone problems	Fish, meat, legumes,grains, drinking water
Iodine	Thyroid hormones synthesis	Enlarged thyroid gland, myxedema, cretinism, increase in blood lipids, gluconeogenesis, and extracelluarretention of NaCl and H_2O	Iodized salt, salt-water sea food, sunflower seeds, mushrooms, liver

Sample questions

1. Define nutrition, nutrients, malnutrition and RDA.
2. Discuss the classification of nutrients in detail.
3. Classify Carbohydrates. Write in detail about the main properties of various classes of carbohydrates.
4. List the functions of protein.
5. Classify lipids and discuss their properties.
6. Enumerate the functions of lipids.
7. What are macro minerals and micro minerals give example?
8. Discuss the functions of iron and calcium?
9. Differentiate between fat soluble and water soluble vitamins?
10. Enumerate the functions of water in our body.

Suggested Readings

Gropper, S.S. and Smith, J.L., 2012. *Advanced Nutrition and Human Metabolism*. Cengage Learning.

Groff, J.L., Gropper, S.S. and Hunt, S.M., 2000. Advanced nutrition and human metabolism. *Belmont (CA): Wadsworth*.

McDowell, L.R., 2003. *Minerals in animal and human nutrition* (No. Ed. 2). Elsevier Science BV.

World Health Organization, 1996. *Trace Elements in Human Nutrition and Health*. World Health Organization.

Davidson, S., Meiklejohn, A.P. and Passmore, R., 1959. Human nutrition and dietetics. *Human nutrition and dietetics.*

Passmore, R. and Eastwood, M.A., 1986. *Davidson and Passmore Human Nutrition and Dietetics: 8th edition continued.* Churchill Livingstone.

Mahan, L.K., Escott-Stump, S. and Krause, M.V., 2007. *Krause's food & nutrition therapy*. Elsevier Saunders.

Bamji, M.S., Krishnaswamy, K. and Brahmam, G.N.V. eds., 2009. *Textbook of Human Nutrition*. Oxford & IBH.

Satyanarayana, S., Dabrowiak, J.C. and Chaires, J.B., 1992. Neither. DELTA.-nor. LAMBDA.-tris (phenanthroline) ruthenium (II) binds to DNA by classical intercalation. *Biochemistry*, *31*(39):9319-9324.

Eastwood, M.A., 2013. *Principles of Human Nutrition*. Springer.

10

Nutrients Requirement of Different Age Groups

Introduction

Nutrition is the fundamental pillar of life and an essential foundation of human and national development. Right from the earliest stage of human development, the fetal stage, at birth, infancy, childhood, adolescence, adulthood and old age, nutrition forms the basis for survival, performance, mental and physical development & economic and social development. Nutritional deficiencies during pregnancy, fetal and early postnatal life can have a negative impact not only on the subsequent growth and development but can also lead to changes in metabolic regulation and homeostasis.

Definitions

Some of the terms used throughout this chapter needs special reference for easy understanding and are hence, discussed below.

Dietary Reference Intake: Dietary Reference Intake (DRI) is the usual term for a batch of reference values adopted to make strategy and determine nutrient intakes by healthy individuals. These values, which changes as per the age and sex, comprises of:

- **Recommended Dietary Allowance (RDA):** It is the average daily intake of nutrients, satisfactory to comply with the requirements of almost all (97% - 98%) healthy individuals.
- **Adequate Intake (AI):** It is initiated when scientific evidenceis inadequate to develop an RDA and is fixed at a level presumed to assure nutritional sufficiency.
- **Tolerable Upper Intake Level (UL):** It is maximum daily intake improbably to induce negative health effects.

Nutritional requirements during life cycle

Before going into the details about the nutritional requirements at different stages of life, one should first know about the different factors considered while calculating and deciding the nutritional requirements. Nutritional needs and concerns vary during different stages of life and for establishing nutrient needs, different facets like populace health, clinical implications from intake levels in both dearth and excess, correlation of disease patterns with each nutrient that requires to be soundly reviewed for all age groups and should be amended time to time are taken into account. For determining the nutritional requirements of individuals or groups, the following factors are kept in mind:

a) The age and sex of the individual or group

b) Body size and composition

c) Physical activity

d) Climate

e) Physical conditions being pregnancy and lactation

The RDA for different nutrients, vitamins, and minerals for Indian population is given by Indian Council for Medical Research (ICMR) and is presented in Table 10.1, 10.2 and 10.3, respectively.

Nutritional requirement during pregnancy

From time immemorial, pregnant and lactating women are considered the most vulnerable section of the society as the nutritional deficiencies in this period will not only have implications on the women but also on the offspring. Thus, it is essential to formulate a nutritionally sufficient diet which will equip the mother with the nutrients necessary for herself and the fetus.

Energy: Energy requirement depends on energy expenditure of the individual which in turn depends on body size and composition in addition to maintenance and physical activity. In pregnant women, energy requirements increase (In first trimester: no ancillary demand for energy; in second trimester, additional demand is 340 kcal/day; in third trimester, additional demand is 452 kcal/day) mainly due to increased maternal body mass, accretion of breast tissue, uterine muscles, placenta, foetal growth, energy needs for increased respiration and heart rate.

Carbohydrates: During pregnancy carbohydrates should contribute approximately 50–60% of total caloric intake. High-carbohydrate foods such as vegetables, fruits, and whole-grain products containing fiber are preferred as they not only protect against constipation but also provide beneficial phytochemicals.

Protein: The recommended protein intake for pregnancy is 1.1g/kg body weight. During pregnancy, protein is required for protein synthesis rather than energy. Protein sources such as chickpeas, lentils, black beans, black-eyed peas, and lima beans and grains like corn, rice, and barley that either complement each other or provide a complete source of protein should be preferred. In case of vegetarian population owing to the low digestibility of vegetable protein, the requirement may be 30% higher than for non-vegetarians.

Fat: Fat serves as a source of fat-soluble vitamins, is primarily used as an energy source and essential fatty acids for foetal growth and development. It is recommended that pregnant women consume 13 g and 1.4 g of the essential fatty acid linoleic acid and alpha-linolenic acid daily, respectively. Safflower, sunflower, corn, and soy oil are rich sources of linoleic acid and flaxseed, soybean, walnut, and canola oils, and leafy green vegetables of alpha-linolenic acid.

Nutrients of Concern

During pregnancy, metabolic demands associated with expansion of maternal tissues and plasma volume, placental, and foetal growth and increased nutrient needs for tissue maintenance increases the requirements for most vitamins and minerals. Most important ones are discussed below.

Folic acid: Folic acid is required in the synthesis of DNA, gene expression, and gene regulation. During pregnancy due to the extensive organ and tissue growth, folate requirements increase dramatically and inadequate folatelevels have been associated with anaemia in pregnancy and reduced foetal growth.

Choline: It is a B-complex vitamin produced by humans and in pregnancy the RDA is 450 mg daily. Its main function is as a precursor of intracellular messengers and as a component of phospholipids in cell membranes.

Vitamin A: It is a key nutrient in pregnancy as it is involved in cell differentiation and its deficiency during early pregnancy can produce malformations of the foetal heart, lungs, and urinary tract.

Vitamin D: Vitamin D in addition to supporting normal functioning of the immune system and inhibiting inflammation supports foetal growth, the addition of calcium to bone, and tooth and enamel formation. Lack of a sufficient supply of vitamin D during pregnancy compromises foetal as well as childhood bone development and can even lead to adverse outcomes like preeclampsia, maternal infection, miscarriage, preterm birth, and the development of asthma and type 1-diabetes in children.

Calcium: It is primarily needed in pregnancy especially in the third trimester for maintenance of maternal bone health and foetal skeletal mineralization. During pregnancy the foetus receives approximately 30g of calcium from the mother.

Iron: During pregnancy the need for iron increases substantially and thus, makes it the most important nutrient. Women require about 1000 mg of additional iron for pregnancy, 450 mg is used to increase red blood cell mass, 250 mg is lost at delivery, and 300 mg is used by the foetus and placenta.

Iodine: During pregnancy iodine is required for thyroid function, energy production, and foetal brain development. Even if mother is deficient in iodine in the early stages of pregnancy, there is an increased risk of hypothyroidism in the offspring.

Nutritional requirement during lactation

Lactation is a remarkably complicated and nutritionally demanding phase in women's lives, as the infant must be fed exclusively on breast milk during the first six months. The infant which grows at a rate of 140 to 200g per week completely relies on the mother for nutrition. The infant needs nutrition not only for the muscle and nerve building which allows the infant to rapidly develop motor skills such as head control and grip but also for the cognitive development. Hence, all of these formative processes require nutritional resources which are provided by mother. This highlights the significance of a healthy diet during lactation period as the nutritional intake of the mother will determine the protein, vitamin, energy, and mineral content of the breast milk.

Energy: For the first six months an additional 500 kcal, and during the next six months an additional 400 kcal is required for a lactating mother. If the extra demand for energy is not met from dietary sources, then the reserved fat stores will be used which can have negative implications on the health of the mother. Energy intake is decided keeping in mind the physical activity of the mother along with the basic goal of maintenance of healthy weight. Maternal diet affects the fatty acid profile and amounts of some vitamins, and trace minerals, whereas, itdoes not significantly alter the protein, carbohydrate, and major mineral composition of breast milk.

Protein: During lactation, there is only a slight increase in protein requirement when compared to that of energy. However, if the protein intake in the mother's diet is not appropriate, then the proportion of casein, an important component which provides calcium and phosphate to the infant may be reduced in breast milk. High-quality protein-rich foods e.g. 175 g of milk or 25 g of cheese or one egg can help in meeting the additional protein requirements during lactation.

Nutrients of Concern

Newborn babies and infants rely on breast milk for an adequate supply of some nutrients (e.g. vitamins A, C, thiamine, riboflavin, B_{12}, B_6, selenium, and iodine). Also, the mothers intake of these nutrients is reflected in the breast milk composition, whereas breast milk composition for nutrients such as folic acid, zinc, iron, vitamin D, calcium, and copper are not affected by mothers intake. The RDA for calcium is 1000 mg/day during the first six months after delivery, as calcium is required for milk production. This requirement can be easily met by adding calcium-rich foods, such as green leafy vegetables and fish in the mother's diet along with 500 ml of milk or milk products per day.

Nutritional requirement of infants

During infancy, the infant experiences rapid growth and development and a healthy new-born doubles its birth weight by 4–6 months and triples it at the end of the first year. However, factors health status, nutritional adequacy, economic and other environmental influences on the family affect the growth of the infant.

Energy: Caloric requirements during infancy depend on growth rate, weight, sleep/wake cycle, climate and temperature, physical activity, metabolic responses to food and health status. Keeping in mind infants rapid growth and development, it becomes imperative to prioritize the energy needs first, because if sufficient calories are not consumed then proteins and carbohydrates will be converted to meet energy needs, thus, over time slowing the growth and development of the infant. Thus, based on growth in breastfed infants, the average caloric need of infants in the first 6 months and 6 to 12 months of age is 108 and 98 Cal/kg body weight.

Protein: Protein requirements during infancy also depend on the same factors as energy; however, as metabolically active muscles need more protein for maintenance, the protein needs are influenced directly by body composition. Recommended protein intake accounts to the average 2.2 g and 1.6 g of protein per kg of body weight from birth up to 6 months and 6 to 12 months, respectively.

Fat: Infants to support their increased requirement for calories, primarily for the functioning of liver, brain, and muscles, including the heart need a concentrated source of energy. Long-chain fatty acids like palmitic (C16:0), stearic (C18:1), and linoleic (C18:2) acids are the most common in food but should be restricted in the infants diet as the young infants are not metabolically equipped to utilized them. Also, infants need cholesterol and its metabolites for brain development and thus, intake should not be restricted in their diet. Essential fatty acids [alpha-linolenic acid (ALA), docosahexaenoic acid (DHA), and eicosapentaenoic acid (EPA)] are essential to build cells, immunity, maintain and regulate the central nervous system, uphold the cardiovascular system, active brain, normal vision plus help the body to acquire nutrients from food.

Nutrients of Concern

During infancy vitamin D and fluoride are considered to be the limiting micronutrients in some environments and hence, supplementation is recommended. Fluoride is required for the enamel of forming teeth, including the not erupted ones. In the deficiency of fluoride the chances of dental caries in early childhood are more frequent; however, tooth discoloration may occur if an infant consumes more fluoride than recommended. Vitamin D is required for bone mineralization along with calcium and hence, for all infants starting shortly after birth 400 IU per day of vitamin D is recommended even when mother's vitamin D dietary needs are complete. In the absence of the supplementation, the infant may develop deficiency disease like rickets which leads to bone deformity for life. Inadequate intake of iodine may induce unusual brain development in the infant which can't be reversed and thus, it is suggested to supplement the diet with it.

Nutritional requirement of preschool and preadolescent children

During toddler (13–36 months) and preschool-age years (3-5 years) growth is slower but steady when compared with infancy, thus resulting in a decreased appetite and food intake in toddlers and pre-schoolers. Protein must be used only for growth and tissue repair and thus, adequate energy intake has to be ascertained to meet an individual child's needs. Also, ingestion of high-quality protein, such as milk and other animal products, should be recommended so that even lower amount of total protein consumed in the diet is able provide the essential amino acids, necessary for the growth and development of the child.

During middle childhood (5-10 years) and preadolescence (9-11 years for girls and 10-12 years for boys) same as the pre-schoolers, the energy needs reflect their slow butsteady growth rate. The energy needs are primarily dependent on child's gender, age, height, weight, and physical activity level. RDA forvarious nutrients for the school-age and preadolescent child have been presented in Table 10.1, 10.2 and 10.3.

Nutritional requirement during adolescence

During adolescence (11-21years) an individual goes through profound biological, emotional, social, and cognitive changes which pave the way to adulthood. However, the intense physical growth and development associated with these changes lead to a significant increase in the nutritional requirements. Additionally, during this period the psychosocial development mainly characterized with the struggle for independence often results in health-compromising eating behaviors, such as meal skipping, excessive dieting, fad diets and use of unconventional nutritional and non- nutritional supplements.

Macronutrients: During puberty, there is a significant increase in lean body mass, skeletal mass, and body fat, resulting in increased energy and nutrient requirements. Energy needs depends onbasal metabolic rate, activity level, and increased requirements to support pubertal growth and development. Also, adolescent males experience greater increases in height, weight, and lean body mass and thus, have significantly higher caloric requirements than females. However, if the energy and protein intakes are consistently not according to the recommended levels then in such individuals the growth may be retarded, delayed sexual maturation and reduced accumulation of lean body mass may be observed. Recent recommendations for the fat consumption suggest that children over the age of 2 years should not consume more than 25–35% of calories from fat, and not more than 10% of calories should be derived from saturated fat.

Nutrients of Concern

During adolescence achieving an adequate intake of calcium, main constituent of bone mass is crucial to physical growth and development and the reduction of the lifetime risk of fractures and osteoporosis. Furthermore, calcium needs and absorption rates during adolescence are higher than any other time except infancy. The increase in blood volume, the rapid rate of linear growth, and the onset of menarche increases an adolescent's need for iron.

Adolescents have increased requirements for folate during puberty as it is an essential part of protein synthesis, DNA, and RNA. Vitamin C is involved in the synthesis of connective tissue and collagen and hence, plays an important role during adolescent growth and development. The RDA for 9- to 13-year-old and 14- to 18- year-old adolescents are shown in Table 10.1, 10.2 and 10.3.

Nutritional requirements during adulthood

Adulthood (20-64 years) is an active growth and development phase, wherein diet, physical activity, smoking, and body weight strongly influence the future course of health and wellness. Recommended total calorie intake from fat, carbohydrate, and protein for adults should be 20–35, 45–65 and 10–35% of calories, respectively. Table 10.1, 10.2 and 10.3enlists recommended nutrient intakes for adults.

Nutrients of Concern

Fiber: Soluble fiber reduces absorption of cholesterol from the gut, thus has positive effects on blood lipid profile, additionally it also helps in glucose metabolism and weight management. It is recommended that on an average 14g of dietary fiber per 1000 calories should be consumed.

Vitamins: Vitamin A is not only important for growth and development of epithelial cells and for activity of immune cells but it also keeps skin and mucous membrane cells healthy and resistant to infection. Several research reports suggest the protective functions of vitamin Din a number of conditions, including cardiovascular disease, hypertension, immune function, cancer and Vitamin E in prostate and esophageal cancer.

Minerals: Calcium plus vitamin D and magnesium not only help in development and maintenance of bone density, which delays osteoporosis and reduces risk of bone fractures, but adequate calcium is also a potential contributor to lowered risk of colon cancer.

Nutritional requirements during old age

Old age rather than a disease is a biological, progressive, inevitable and irreversible phenomenon. As people age, physiological changes occur in most of the organs along with a concomitant increase in the presence and number of chronic conditions such as cardiovascular disease, hypertension, diabetes, osteoporosis, and dementia. Monitoring the nutritional requirements as per the need not only influences present health but has also emerged as a major modifiable determinant of chronic disease and age-related decline.

Energy: Energy requirements primarily depend on basal metabolic rate (slows with age), diet-induced thermo genesis, and physical activity, along with genetics, hormones, and body composition which influences metabolic rates and thus, result in broad ranges of energy needs for older populations. Although the energy requirements decrease as the age increases but the nutrient density remains same.

Carbohydrate and Fiber: Generally, carbohydrate intake of an individual should meet approximately 45 to 65% of calories and depending on caloric intake, daily 14 g/1000 calories of dietary fiber is recommended foradults. Dietary fiber not only reduces the risk of coronary heart disease but also helps in maintaining the gastrointestinal health.

Protein: The RDA for protein for old age people is 0.8 g per kg body weight/ day, also it is observed that old people follow the low-calorie diet and thus, are often protein deficient. Old age is characterized with decreasing muscle mass which does not lead to lower protein requirements, but in contrast, such an individual requires higher protein intake. Consuming enough protein as recommended in the RDA not only stimulates muscle synthesis but also minimizes the risk of muscle wasting (sarcopenia) in aging adults.

Fat: The recommendations for dietary fat remains the same as for other age groups, i.e. the amount of saturated fat should be minimized in the diet to maintain a beneficial blood cholesterol ratio, high saturated fat and trans fatty acid should be avoided to prevent the risk of various cardiovascular diseases.

Nutrients of Concern

Older adults suffer invariably form age-associated metabolic changes or low dietary intake and thus, the nutrients discussed in the following sections are of special concern for this age group.

Vitamins: Due to decreased clearance of vitamin A metabolites (retinyl esters) from the blood or due to increased absorption, the plasma levels and liver stores of vitamin A increase with age. Also, sometimes in individuals suffering from kidney disease the retinol-binding protein (vitamin A metabolite) is not cleared from blood which further elevates serum vitamin A levels. These factors make older adults vulnerable to vitamin A toxicity and possible liver damage. Thus, care should be taken while including Vitamin A in the diet or while taking supplements of the same.

Vitamin E, the antioxidant vitamin plays a significant role in the health of older adults due to its antioxidant functions, such as hindering development of cataracts, along with enhanced immune function and cognitive status. Interest in vitamin K supplementation in old age is increasing owing to its role in blood coagulation, reducing the risk of bone fractures, in addition to the presence of vitamin K-dependent proteins in bone, vascular, and centralnervous system tissues. It is estimated that 30% of old age people suffer from atrophic gastritis which decreases the levels of B_{12} in blood and causes irreversible neurological symptoms which include deterioration of mental function, change in personality and loss of physical coordination are observed.

Minerals: Research studies suggest that low calcium intake have direct/indirect effect on colon cancer, overweight, and hypertension, hence, nutritional recommendations should be followed for calcium intake.

Magnesium is essential for bone and tooth formation, nerve activity, glucose utilization, and synthesis of fat and proteins. Old age people suffering from chronic alcoholism, malabsorption due to gastrointestinal disorders, and diabetes etc. can develop its deficiency. Signs of deficiency include muscle spasms, weakness, irritability, aggressiveness, vertigo, and seizures. Also, drugs such as magnesium hydroxide or citrate laxatives may lead to magnesium overdose, which can lead todiarrhea, dehydration, and impaired nerve activity.

Table 10.1: RDA of nutrients in different age groups

Life Stage	Total (l/d)	Protein (g/d)	Carbohydrate (g/d)	Fat (g/d) (g/d)	Dietary fibre	Essential water fatty acids (mg/d)
Infants and children						
0–6 m	0.6	6.26	60	31	-	0.5
7–12	0.8	14.2	95	30	-	0.5
1–3 y	1.3	16.7	130	-	19	0.7
4-6 y	-	20.1	-	-	-	-
4–8 y	1.7	-	130	-	25	0.9
7-9 y	-	29.5	-	-	-	-
Older Children (Males)						
9–13 y	2.4	-	130	-	31	1.2
10-12 y	-	39.9	-	-	-	-
13-15 y	-	54.3	-	-	-	-
16-18 y	-	61.5	-	-	-	-
14–18 y	3.3	-	130	-	38	1.6
Adults (Males)						
19–50 y	3.7	-	130	-	38	1.6
>50 y	3.7	-	130	-	30	1.6
Sedentary work	-	60	-	-	-	-
Moderate work	-	60	-	-	-	-
Heavy work	-	60	-	-	-	-
Older children (Females)						
9–13 y	2.1	-	130	-	26	1.0
10-12y	-	40.4	-	-	-	-
13-15 y	-	51.9	-	-	-	-
16-18 y	-	55.5	-	-	-	-
14–18 y	2.3	-	130	-	26	1.1
Adults (Females)						
19–50 y	2.7	-	130	-	25	1.1
>50 y	2.7	-	130	-	21	1.1
Sedentary work	-	55	-	-	-	-
Moderate work	-	55	-	-	-	-
Heavy work	-	55	-	-	-	-
Pregnancy(19–50 y)	3.0	78	175	-	28	1.4
Lactation (0-6 m)	3.8	74	210	-	29	1.3
Lactation (6-12 m)	3.8	68	210	-	29	1.3

Source: NIN (2010)

Table 10.2: Recommended dietary allowances for vitamins

Life Stage	Vitamin A (mg/d)	Vitamin D(mg/d)	Vitamin E(mg/d)	Vitamin K(mg/d)	Vitamin B_1(mg/d)	Vitamin B_2(mg/d)	Vitamin B_3(mg/d)	Vitamin B_5(mg/d)	Vitamin B_6(mg/d)	Vitamin B_7(ìg/d)	Vitamin B_9 (mg/d)	Vitamin B_{12} (ìg/d)	Vitamin C (mg/d)
Infants and children													
0–6 m	350.0	5.0	4.0	2.0	0.2	0.3	4.0	1.7	0.1	5.0	25.0	0.2	25.0
7–12	350.0	5.0	5.0	2.5	0.3	0.4	6.0	1.8	0.4	6.0	25.0	0.2	25.0
1–3 y	400.0	5.0	6.0	30.0	0.5	0.6	8.0	2.0	0.9	8.0	30.0	1.0	40.0
4-6 y	400.0	-	-	-	0.7	0.8	11.0	-	0.9	-	40.0	1.0	40.0
4–8 y	-	5.0	7.0	55.0	-	-	-	3.0	-	12.0	-	-	-
7-9 y	400.0	-	-	-	0.8	1.0	13.0	-	1.6	-	60.0	1.0	40.0
Older Children (Males)													
9–13 y	-	5.0	11.0	60.0	-	-	-	4.0	-	20.0	-	-	-
10-12 y	600.0	-	-	-	1.1	1.3	15.0	-	1.6	-	70.0	1.0	40.0
13-15 y	600.0	-	-	-	1.4	1.6	16.0	-	2.0	-	100.0	1.0	40.0
16-18 y	600.0	-	-	-	1.5	1.8	17.0	-	2.0	-	100.0	1.0	40.0
14–18 y	-	5.0	15.0	75.0	-	-	-	5.0	-	25.0	-	-	-
Adults (Males)													
19–50 y	-	5.0	15.0	120.0	-	-	-	5.0	-	30.0	-	-	-
>50 y	-	10.0	15.0	120.0	-	-	-	5.0	-	30.0	-	-	-
Sedentary work	600.0	-	-	-	1.2	1.4	16.0	-	2.0	-	100.0	1.0	40.0
Moderate work	600.0	-	-	-	1.4	1.6	18.0	-	2.0	-	100.0	1.0	40.0
Heavy work	600.0	-	-	-	1.7	2.1	21.0	-	2.0	-	100.0	1.0	40.0
Older children (Female)													
9–13 y	-	5.0	11.0	60.0	-	-	-	4.0	-	20.0	-	-	-
10-12y	600.0	-	-	-	1.0	1.2	13.0	-	1.6	-	70.0	1.0	40.0
13-15 y	600.0	-	-	-	1.2	1.4	14.0	-	2.0	-	100.0	1.0	40.0
16-18 y	600.0	-	-	-	1.0	1.2	14.0	-	2.0	-	100.0	1.0	40.0
14–18 y	-	5.0	15.0	75.0	-	-	-	5.0	-	25.0	-	-	-

(Contd.)

Adults (Female)													
19–50 y	-	5.0	15.0	90.0	-	-	-	5.0	-	30.0	-	-	-
>50 y	-	10.0	15.0	90.0	-	-	-	5.0	-	30.0	-	-	-
Sedentary work	600.0	-	-	-	1.0	1.1	12.0	-	2.0	-	100.0	1.0	40.0
Moderate work	600.0	-	-	-	1.1	1.3	14.0	-	2.0	-	100.0	1.0	40.0
Heavy work	600.0	-	-	-	1.4	1.7	16.0	-	2.0	-	100.0	1.0	40.0
Pregnancy(19–50 y)	800.0	5.0	15.0	90.0	1.2	1.5	14.0	6.0	2.5	30.0	400.0	1.2	60.0
Lactation (0-6 m)	950.0	5.0	19.0	90.0	1.3	1.4	16.0	7.0	2.5	35.0	150.0	1.5	80.0
Lactation (6-12 m)	950.0	5.0	19.0	90.0	1.2	1.5	15.0	7.0	2.5	35.0	150.0	1.5	80.0

Source: NIN (2010)

Table: 10.3: Recommended dietary allowances for minerals for different age groups

Life Stage	Calcium (mg/d)	Chromium (ig/d)	Copper (ig/d)	Fluoride (mg/d)	Iodine (ig/d)	Iron (mg/d)	Magnesium (mg/d)	Phosphorus (mg/d)	Zinc (mg/d)	Potassium (mg/d)	Sodium (g/d)	Chloride (g/d)
Infants and children												
0–6 m	500	0.2	200	>0.01	90	0	30	750	0	350-925	0.12	0.18
7–12	500	5.5	220	0.5	90	5	45	750	0	475-1275	0.37	0.57
1–3 y	600	11	340	0.7	-	9	50	600	5	550-1650	1.0	1.5
1-5 y		-	-	-	90	-	-	-	-	-	-	-
4-6 y	600	-	-	-	-	13	70	600	7	775-2325	-	-
4–8 y	-	15	440	1	-	-	-	-	-	-	1.2	1.9
7-9 y	600	-	-	-	-	16	100	600	8	-	-	-
6-12 y	-	120	-									
Older Children (Males)												
>13y	-	-	-	-	150	-	-	-	-	-	-	-
9–13 y	-	25	700	2	-	-	-	-	-	-	1.5	2.3
10-12 y	800	-	-	-	-	21	120	800	9	-	-	-
13-15 y	800	-	-	-	-	32	165	800	11	-	-	-
16-18 y	800	-	-	-	-	28	195	800	12	-	-	-
14–18 y	-	35	890	3	-	-	-	-	-	-	1.5	2.3
Adults (Males)												
19–50 y	-	35	900	4	-	-	-	-	-	3750	1.5	2.3
>50 y	-	30	900	4	-	-	-	-	-	-	1.3	2.3
Sedentary work	600	-	-	-	-	17	340	600	12	-	-	-
Moderate work	600	-	-	-	-	17	340	600	12	-	-	-
Heavy work	600	-	-	-	-	17	340	600	12	-	-	-

(Contd.)

Older children (Female)												
9–13 y	-	21	700	2	-	-	-	-	-	-	1.5	2.3
10-12y	800	-	-	-	-	27	160	800	9	-	-	-
13-15 y	800	-	-	-	-	27	210	800	11	-	-	-
16-18 y	800	-	-	-	-	26	235	800	12	-	-	-
14–18 y	-	24	890	3	-	-	-	-	-	-	1.5	2.3
Adults (Female)												
19–50 y	-	25	900	3	-	-	-	-	-	3225	1.5	2.0
>50 y	-	20	900	3	-	-	-	-	-	-	1.3	2.0
Sedentary work	600	-	-	-	-	21	310	600	10	-	-	-
Moderate work	600	-	-	-	-	21	310	600	10	-	-	-
Heavy work	600	-	-	-	-	21	310	600	10	-	-	-
Pregnancy(19–50 y)	1200	30	1000	3	250	35	310	1200	12	12000	1.5	2.3
Lactation (0-6 m)	1200	45	1300	3	250	21	310	1200	12	-	1.5	2.3
Lactation (6-12 m)	1200	45	1300	3	250	21	310	1200	12	-	1.5	2.3
Postmenopausal								600-800				

Source: NIN (2010)

Sample questions

1. What is nutrition? How it differs according to age?
2. Give a brief account of nutritional requirement during pregnancy.
3. What is the nutritional requirement of the lactating mother?
4. Discuss in brief nutritional requirement of infants and pre-school children.
5. Briefly discuss the nutritional requirement of school children and teenagers.
6. Give a brief account of nutritional requirement of adults.
7. Discuss in brief the nutritional requirement in old age.
8. How the energy requirement of women changes during pregnancy?
9. What is the importance of energy requirement in lactating mother?
10. Give a brief account of body energy needs of all age groups.
11. Briefly explain RDA, AI, UL,and PAL.
12. What are the factors affecting the nutritional requirement of elderly people?
13. How the nutritional and energy requirement of pregnant and lactating women calculated?
14. Why is the energy need higher in school children and teenagers?
15. Discuss in brief RDA for nutrients in different age groups.
16. How physical activity of a person affects the nutritional and energy requirement?
17. What nutrient are prerequisites for fetus growth and development?

Suggested Readings

A report of a joint FAO/WHO expert consultation, (1998). *Vitamin and mineral requirements in human nutrition*. Bangkok, Thailand.

Burgess, A. and Peter, G. (2004). *Family nutrition guide*. Food and Agriculture Organization of the United Nations. Rome.

Desai B. B. (2000). *Handbook of Nutrition and diet*. New York: Marcel & Dekker Inc.

Eskin, N. M. A. and Shahidi, F. (2013). *Biochemistry of foods*. 3rdedn. USA: Academic Press Elsevier.

Food and Nutrition Board, Institute of Medicine. (2005). *Dietary Reference Intakes for Energy, Carbohydrate, Fiber, Fat, Fatty Acids, Cholesterol, Protein, and Amino Acids*. Washington DC: National Academies Press. Available at www.nap.eduhttps://www.nal.usda.gov/fnic/dri-nutrient-reports.

Leslie, W. and Hankey, C. (2015). Aging, Nutritional Status, and Health. *Healthcare,* 3, pp. 648-658.

Linder, M. C. (1985). *Nutritional biochemistry and metabolisms with clinical applications*. New York: Elsevier Science publishing company Inc.

Ministry of Health, (2013). *Food and Nutrition Guidelines for Healthy Older People: A background paper*. Wellington.

Park, Y. W. and Haenlein, G. W. F. (2013). *Milk and Dairy Products in Human Nutrition (Production, composition, and health)*. UK: Wiley Blackwell.

Rajalakshmi, R. Rev. by Sakariah, K. K. (2013). *Applied nutrition*. 4thedn. New Delhi: Oxford & IBH Publishing Co. Pvt. Ltd.

Sardesai, V. (2012). *Introduction to clinical nutrition*. 3rd ed. USA: CRC press (Taylor & Francis Group).

Srilakshmi, B. (2009). *Human Nutrition (for B.Sc. Nursing students)*. New Delhi: New age International (P) Ltd., Publishers.

Talwar G.P., Srivastava L.M., and Moudgil, K.D. (1989). *Textbook of biochemistry and human biology*, 2ndedn. New Delhi: Prentice Hall of India Pvt. Ltd.

United Nations. (2017). *World Population Aging Report*. Available at http://www.un.org/en/development/desa/population/publications/pdf/ageing/WPA2017_Highlights.pdf

11

Evaluation of Nutritive Value of Foods

Introduction

Nutritive value of any food indicates the calories content as well as quality attributes. Macronutrients such as carbohydrates, fats and proteins provide energy by 4-9-4 rule. It means they provide 4, 9 and 4 kcal/g energy, respectively, when oxidized completely. Though micronutrients such as vitamins and minerals do not provide any calories, they are essential for the enzymatic activity and metabolic reactions.

ATP or energy is measured in terms of 'calories' or 'joule'. One calorie is the amount of heat required by 1g of water to increase its temperature from 14.5 to 15.5°C under standard conditions. Nutritive value of any food can be measured on shelf using calorimeter, in which food is completely oxidized under specified conditions and converted into CO_2 and water.

$$1 \text{ kcal} = 1 \text{ Cal} = 1000 \text{ calories}$$

Measurement of Calories Content of Food

Calories content of any food can be measured by different calorimeter. Any insulated/jacketed device which can measure the amount of heat absorbed or released during a chemical reaction is known as 'calorimeter'. In order to determine the calorie content of food, the most commonly used calorimeter is Bomb calorimeter. Bomb calorimeter is a constant-volume calorimeter, containing a vessel of steel fully jacketed, equipped with two electrodes to oxidize the food or any substance placed in it. It is constructed to withstand the pressure produced by the reaction (oxidation) as it heats the air within the container. The temperature change of water is used to calculate the heat of combustion and total heat change of the system. A schematic diagram of bomb calorimeter is shown below (Fig. 11.1):

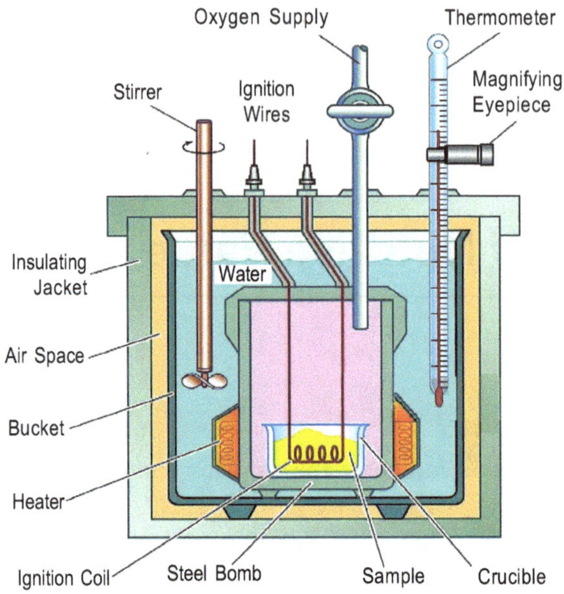

(*Source*: Bomb calorimeter, Getty images)
Fig. 11.1: A schematic diagram of Bomb Calorimeter

Bomb is a vessel made up of steel in such a way so that it can withstand the high pressure generated during expansion of gases. It also has valves for the entry of ignition wires (electrodes) and gases (oxygen).A known amount of sample is placed in a crucible and that crucible is kept in a closed steel vessel (bomb). This bomb vessel is surrounded by a known volume of water, which is further closed by an insulated jacket vessel to prevent the transfer of heat and gases from outside to inside or inside to outside.

By using electrodes, complete combustion (oxidation) of sample takes place in presence of sufficient oxygen; and CO_2, water and heat is generated as a result of oxidation. As the bomb vessel can withstand the high pressure; generated heat due to oxidation is not utilized in the expansion of vessel or pressure-volume work; and all heat is transferred to the surrounding air to increase its temperature. As again, the water is surrounded by insulated jacket, no heat is transferred to outside and thus; change in temperature of water is measured. Here, in bomb calorimeter, heat released due to chemical reaction (oxidation) is not measured as change of enthalpy (ΔH), but it is measured as change of internal energy (ΔU).

By using reference (benzoic acid) of which heat combustion is already known, the change of internal energy can be calculated for any unknown substance. The ΔU can be converted into ΔH by the following reaction:

$$\Delta H = \Delta E + \Delta n\ (RT)$$

Where, Δn indicates the change in number of moles of gas during oxidation, R indicates gas constant (0.008314 $kJK^{-1}mol^{-1}$) and T represents the average water temperature (K).

Important: Now-a-days, bomb calorimeter is not used for the estimation of energy/ calorie content of foods. Instead, 'Atwater system' named after Wilbur Olin Atwater, is used for the calculation of available energy of any food. In this system, total caloric value is calculated by adding up the calories provided by the energy-containing nutrients: protein, carbohydrate, fat and alcohol. Industries use the simple calculation method of 4-9-4 rule, in which the amount of carbohydrate, fats and protein is measured and energy is calculated accordingly. To calculate energy from alcohol, the factor of 7 kcal/g is used (1g alcohol = 7 kcal energy). As food also contains dietary fibers which are not digested by human body, they are subtracted from the total carbohydrates to calculate the correct energy value of food. In next sections, we will study how to calculate the basic composition i.e. carbohydrate, proteins and fats in any food quantitatively as well as qualitatively.

Estimation of carbohydrates

For the determination of nutritive value of foods, amount of carbohydrates needs to be quantified. The carbohydrate content can be determined by calculating the percent remaining after all the other components have been measured as follows:

$$\%\ \text{carbohydrates} = 100 - [\%\ \text{moisture} + \%\ \text{protein} + \%\ \text{lipid} + \%\ \text{mineral}]$$

Although this method is most commonly used for the quantification of carbohydrates, but sometimes it gives erroneous results due to possible manual error in estimation of moisture or other components. Carbohydrates can be measured quantitatively by the following methods:

Quantitative tests for carbohydrates

Different tests are given for the quantitative estimation of carbohydrates in Table 11.1.

Table 11.1: Tests for quantitative estimation of carbohydrates

Type of method	Name of the method/test	Specific for
Titration method	Lane-Eynon method	Reducing sugars
Gravimetric method	Munson and Walker method	Reducing sugars
Colorimetric method	Anthrone method	Total sugars
Colorimetric method	Phenol - Sulfuric Acid method	Total carbohydrates
Physical method	Polarimeter	Type of sugars
Physical method	Refractive index	Total sugars
Enzymatic method	Glucose oxidase method	Glucose

(*Source*: http://people.umass.edu/~mcclemen/581Carbohydrates.html)

The brief of some of the tests are given below. The details of the above mentioned tests can be found at the source given above.

1. Lane-Eynon method

Unknown solution (containing carbohydrates) is added drop-wise using a burette into a flask containing boiling alkaline copper sulphate solution added with methylene blue indicator. Initially during titration, cupric ions (Cu^{2+}) will be reduced to cuprous (Cu_2^{2+}) ions and give bright-orange color. Now, dilute the mixture of flask with water in 1:1 ratio and repeat the titration.

Do a second titration by adding all but 1 ml of the titre (sample) solution at once and boil the mixture for 2 min. Add 3-4 drops of methylene blue indicator and complete the titration until color change from blue to colorless. Note the titre value and calculate the percent reducing sugars.

2. Anthrone test

This test indicates the total sugar present in the sample. Under highly acidic conditions, reducing or non-reducing sugars produce furfurals, which give bluish-green complex on reaction with anthrone reagent. This colored complex can be measured calorimetrically or by spectrophotometer at 620nm. By using calibration curve, the concentration of total sugars can be determined. Under highly acidic conditions, all non-reducing sugars are converted into reducing sugars, and thus, this method gives the total sugars of the sample. The reaction is as shown below (Fig. 11.2):

3. Phenol - Sulfuric Acid method

This method indicates the total carbohydrates/total sugars present in the sample. Unknown sample is mixed with phenol under highly acidic conditions. Sugars react with phenol and develop a yellow-orange colored complex, which can be measured spectrophotometrically at 420nm.

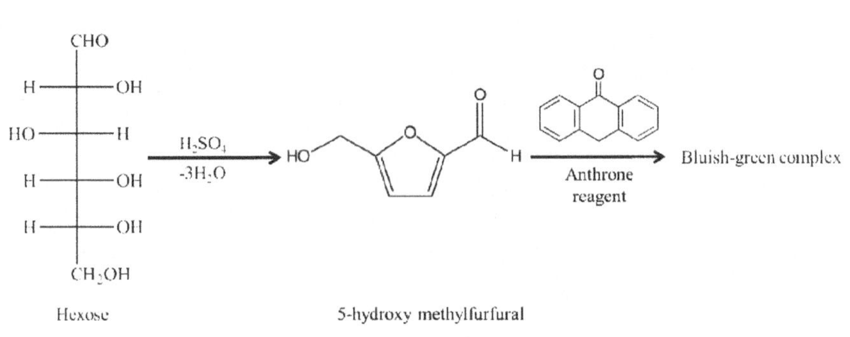

Fig. 11.2: Chemical reaction for Anthrone test

Qualitative tests for carbohydrates

Chemical method	Specific for
Molisch's Test	Carbohydrates
Fehling's Test	Reducing sugars and aldehydes
Barfoed's Test	Reducing sugars (mainly monosaccharides)
Benedict's Test	Reducing sugars
Seliwanoff's Test	Ketoses
Bial's Test	Pentoses
Iodine Test	Starch/polysacchrides

1. Molisch's Test

Sugar solution is added with sulphuric acid and produce dehydrated products (furfurals). On addition of 2-3 drops of beta-naphthol solution, a purple colored ring in test tube indicates the presence of sugar. The reaction is shown below (Fig. 11.3):

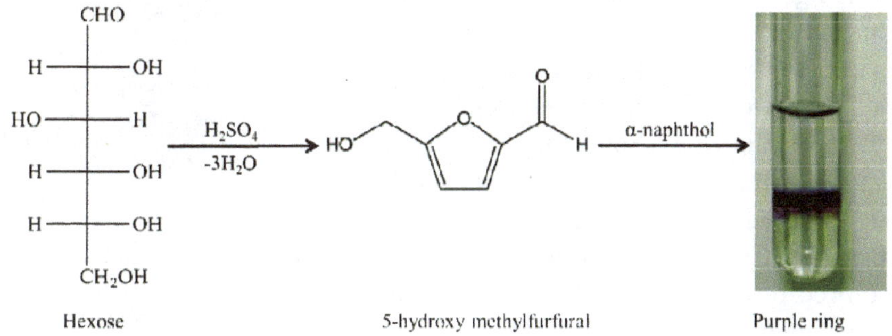

Fig. 11.3: Chemical reaction for Molisch's test

2. Fehling's Test

When the sample solution is added with Fehling solution in equal amount and boiled for 10 min, a red color precipitate indicates the presence of reducing sugars. During reaction, cupric ions (Cu^{2+}) are reduced by sugar into brick-red colored cuprous ions (Cu_2^{2+}). [Fehling solution is the mixture of copper sulphate and alkaline solution of sodium potassium tartrate (Rochelle salt)]. The reaction is as shown below (Fig. 11.4):

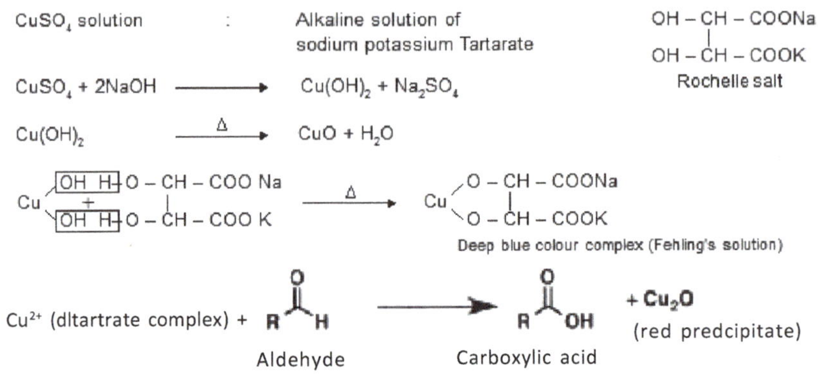

Fig. 11.4: Chemical reaction for Fehling's test

3. Barfoed's Test

Barfoed's reagent, cupric acetate in acetic acid, is slightly acidic and can be reduced only by powerful reducing agents. Disaccharides may also react with this reagent, but the reaction is much slower when compared to monosaccharides. Sample solution is reacted with freshly prepared Barfoed reagent and heated for 3 min in boiling water bath followed by cooling. During reaction, cupric ions (Cu^{2+}) get reduced by reducing sugar into brick-red colored cuprous ions (Cu_2^{2+}). Blue colored solution with brick red colored precipitates settled at bottom indicates the presence of reducing sugars (Fig. 11.5).

Fig. 11.5: Positive and negative indication for Barfoed's Test.

4. Benedict's test

Benedict's reagent is a complex mixture of sodium carbonate, sodium citrate and copper (II) sulfate pentahydrate. Sample solution is reacted with Benedict's reagent and heated for 5 min in boiling water bath followed by cooling. During reaction, cupric ions (Cu^{2+}) get reduce by reducing sugar into to cuprous ions (Cu_2^{2+}), color of which depends on the concentration of reducing sugar. Possible colors of final reaction mixture can be obtained as below (Fig. 11.6):

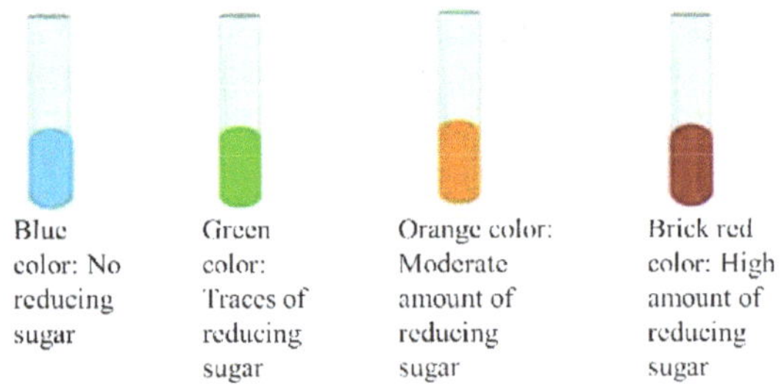

Fig. 11.6: Interpretation of Benedict's test results

5. Seliwanoff's Test

This reagent consists of resorcinol and concentrated hydrochloric acid. When a few drops of sample solution are mixed with Seliwanoff reagent and heated to just boiling; appearance of cherry red condensation product will indicate the presence of ketoses (e.g. fructose) in test sample (Fig. 11.7). There will be no significant change in colour for aldose (e.g. glucose) sugar.

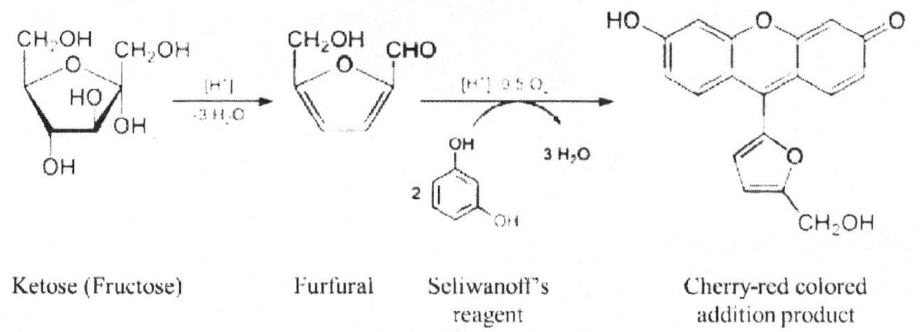

Fig. 11.7: Chemical reaction for Seliwanoff's Test

6. Bial's Test

Bial's reagent includes orcinol, hydrochloric acid, and ferric chloride. When Bial reagent reacts with sample solution under boiling water bath for 2 min, bluish-green colored product indicates the presence of pentoses (5-C sugars) (Fig. 11.8). Hexoses generally react to form muddy brown products.

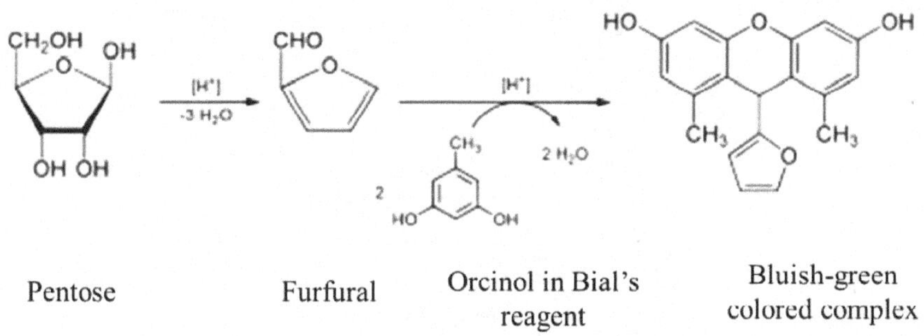

Fig. 11.8: Chemical reaction for Bial's test

7. Iodine test

When iodine solution reacts with test solution and give blue-black colored solution, it indicates the presence of polysaccharides (starch) (Fig. 11.9).

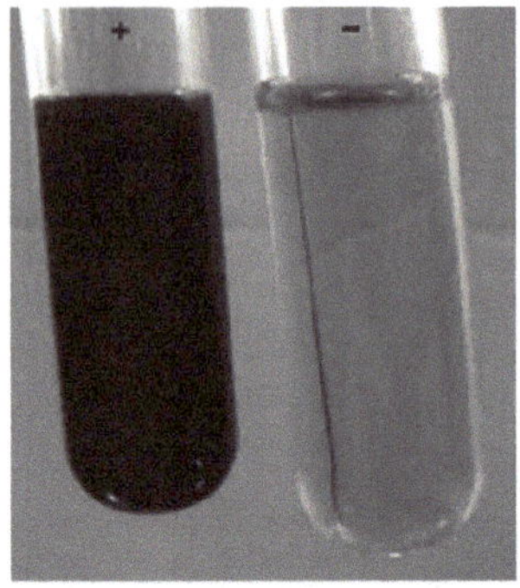

Positive Negative

Fig. 11.9: Color indication for positive and negative iodine test

Estimation of proteins

Proteins can be estimated quantitatively by Kjeldahl method and qualitatively by several methods, which are explained below:

1. Kjeldahl method

Kjeldahl method is the oldest and most commonly used for the quantitative estimation of proteins in food. This method measures the amount of nitrogen, which is then multiplied by a factor (~6.25) to get protein content. Any protein contains approximately 16% protein, therefore, 100/16 = 6.25 factor is used for the multiplication with nitrogen concentration to get protein content.

Kjeldahl method involves three steps:

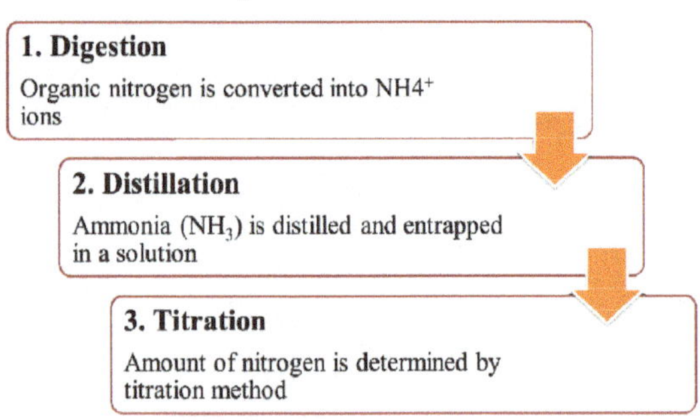

i. Digestion: Food sample is digested (and oxidized) in presence of highly acidic environment to convert it into CO_2 and water. Proteins are converted into ammonia gas (NH_3). In this step, sulphuric acid acts as oxidizing agent and anhydrous sodium sulphate or potassium sulphate increases the boiling point of the reaction mixture (up to 300-350°C). Higher temperature facilitates the digestion process. A catalyst such as copper, selenium, titanium, or mercury is also used, which increases the rate of the reaction. Digestion process continues till a clear solution is obtained. It may take 8-10 hours. Produced NH_3gas is not escaped through acidic solution as it forms ammonium sulphate [$(NH_4)_2SO_4$] by reacting with sulphate ions.

$$\text{Protein (-N)} + H_2SO_4 \xrightarrow{\text{Catalyst \& Sodium sulphate}} (NH_4)_2SO_4 + CO_2 + H_2O$$

ii. Distillation: After completion of digestion, excess of sodium hydroxide is added into it to make the solution alkaline. Now, ammonium ions are converted into ammonia which is liberated from digestion flask and collected in another flask containing boric acid solution through distillation. Boric acid is converted into borate ions and ammonia gas is again converted into ammonium ions/ ammonium borate

$$(NH_4)_2SO_4 + 2NaOH \rightarrow 2NH_3 + 2H_2O + Na_2SO_4$$

$$NH_3 + H_3BO_3 \text{ (boric acid)} \rightarrow NH_4^+ + H_2BO_3^- \text{(borate ion)}$$

iii. Titration: The concentration of captured ammonium ions can be determined by simply titrating it with any acidic solution such as standard sulphuric acid or HCl using suitable indicator. Thus, the amount of nitrogen can be obtained by using amount of acid used in titration and suitable formula.

2. Biuret test

The Biuret reaction can be used for both qualitative and quantitative analysis of protein. The biuret method depends on the presence of peptides bonds in proteins. When a solution of proteins is treated with cupric ions (Cu^{2+}) in a moderately alkaline medium, a purple colored Cu^{2+}- peptide complex is formed which can be measured quantitatively by spectrophotometer in the visible region (540nm) (Fig. 11.10).The intensity of the color produced is proportional to the number of peptide bonds that are reacting, and therefore, to the number of protein molecules present in the reaction system. The reaction neither occurs with free amino acids due to absence of peptide bonds nor with dipeptides due to presence of only one peptide bond. Biuret reaction needs presence of at least two peptide bonds in a molecule. Therefore, it gives positive test with tri-, oligo- and polypeptides.

Assaying for proteins using this method does not actually use biuret. Biuret is a small compound that forms when urea is heated which causes two urea molecules to join. Urea molecules fused in this manner produce amide groups (-NH) at the center of the molecule which bind to copper ions at a basic pH. The copper complexes that result from this interaction produce a strong blue color that can be measured with a spectrophotometer.

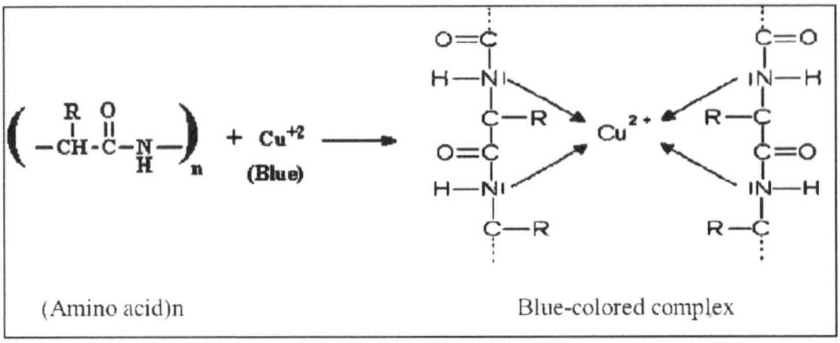

Fig. 11.10: Chemical reaction for biurete test

3. Ninhydrin test

Ninhydrin (2,2-dihydroxyindane-1,3-dione) test is used to detect amino acid or proteins containing free amino group. Almost all amino acids (except proline) give positive test with ninhydrin. Ninhydrin is a strong oxidizing agent and causes deamination of amino acid and liberates free ammonia. In presence of liberated ammonia, ninhydrin gives the purple-blue colored 'Ruhemanns complex' (diketohydrin) (Fig. 11.11).

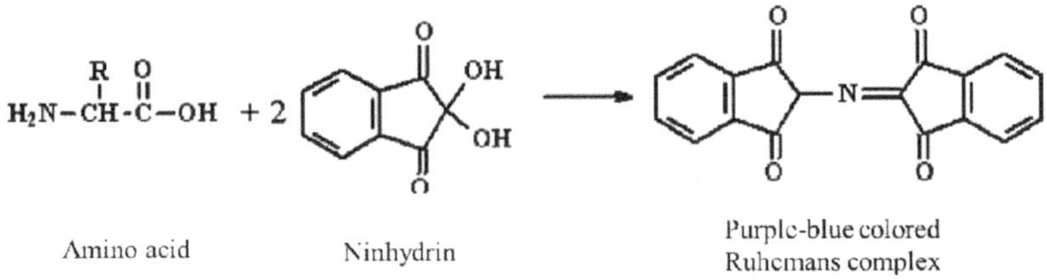

Fig. 11.11: Chemical reaction for Ninhydrin test

4. Lowry's method

This method indicates the protein content quantitatively and involves Lowry reagent. This reagent is the mixture of copper sulphate, sodium potassium tartarate and sodium carbonate. Lowry method combines biuret reagent with Folin-Ciocalteau reagent. Folin-Ciocalteau (F.C.) reagent consists of sodium tungstate molybdate and phosphate. In presence of Lowry reagent and F.C. reagent, protein solution gives deep blue color, which can be measured by spectrophotometer between 500-750nm. Using a standard curve, concentration of test sample can be calculated.

5. Xanthoproteic Test

Xanthoproteic test is used to detect aromatic amino acids (tyrosine, tryptophan and phenylalanine) in a protein solution. When test sample is heated in presence of conc. nitric acid, yellow colored nitro derivatives of amino acid are formed. Yellow color indicates the presence of aromatic amino acids (Fig. 11.12).

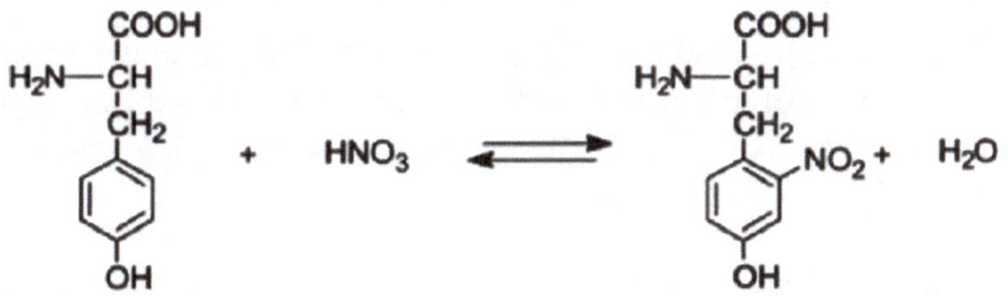

Tyrosine

Nitro derivative
(Yellow colored)

Fig. 11.12: Chemical reaction for Xanthoproteic test

6. Millon's test (Cole's test)

This test is specific for amino acid with phenolic structure, i.e. tyrosine. Millon's reagent is acidified mercuric sulphate. When tyrosine (free or in protein) reacts with Millon's reagent, it gives yellow precipitate of mercury-protein complex. Further, on addition of sodium nitrite solution and heating, the yellow colored complex is converted into mercury phenolate which is red in colour (Fig. 11.13 and 11.14).

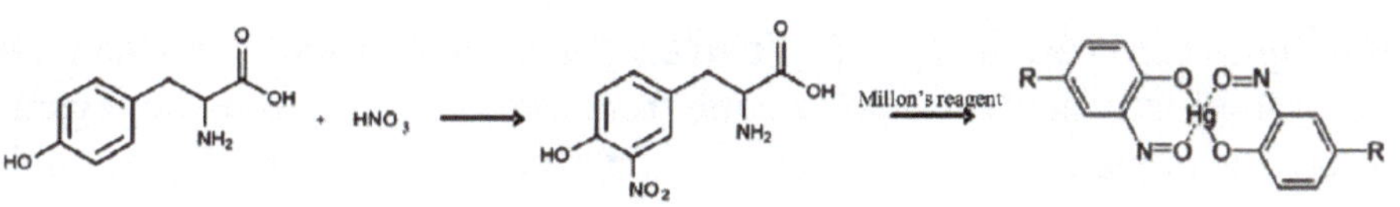

Tyrosine

Nitrated tyrosine

Red colored mercury phenolate

Fig. 11.13: Chemical reaction for Millon's test

Million's test Positive

Fig. 11.14: Color indication for positive Millon's test

7. Sakaguchi Test

Sakaguchi reagent consists of α-naphthol and sodium hypobromite. This test is specific for detection of amino acid containing gauanidium group, e.g. arginine. In the presence of an oxidizing agent like bromine solution, it gives red color with α-naphthol.

8. Nitroprusside Test

This test is specific for the detection of free thiol group/sulphydryl group in amino acid, which is present only in cysteine. In presence of nitroprusside, free sulphydryl group of amino acid gives red colored complex, indicating the presence of cysteine in protein (Fig. 11.15).

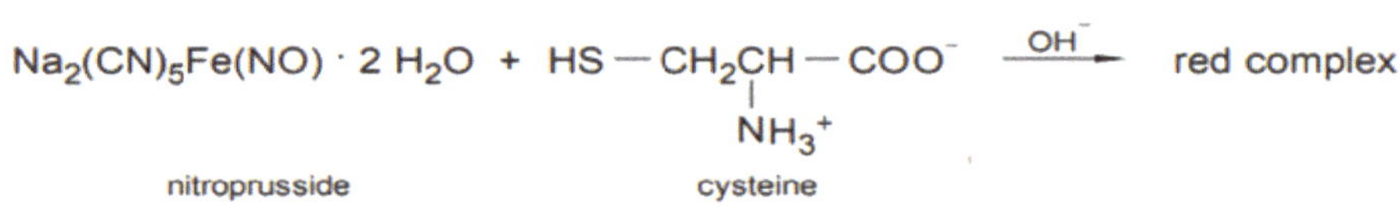

Fig. 11.15: Chemical reaction for Nitroprusside test

Estimation of lipids

Fats or lipids can be estimated quantitatively (total content) and/or qualitatively (type of lipid/fatty acid) after extracting from the food using suitable organic solvent. The property of fats of being soluble in organic solvents and insoluble in water make the organic solvents the best extraction medium of fats from foods. The amount of lipid extracted from any food sample depends on the nature of organic solvent. A few organic solvents are polar in nature such as ethanol, methanol, etc., which are suitable for the extraction of polar lipids like glycolipids, phospholipids, etc. However, most of the solvents are non-polar solvents such as petroleum ether, diethyl ether which are suitable for the extraction of non-polar lipids like triacylglycerol. As food lipid consists of both type of components (polar as well as non-polar lipids); therefore, a single solvent would not be able to extract complete fat components from food and thus, a mixture of organic solvents will be required for the extraction of lipids. Fats/ lipids can also be extracted from foods without using any solvent (non-solvent methods) or by using supercritical fluid (CO_2), which will also be discussed in brief.

Solvent extraction methods

Before extraction of fat using solvents, few preparatory steps are followed:

1. It is preferable to dry the sample before extraction of fat as it is difficult to penetrate in foods containing water for organic solvents resulting incomplete extraction of fat.

2. Sample size should be reduced to increase the surface area and action sites for organic solvent.

3. Some foods contain lipids complexed with proteins (lipoproteins) and carbohydrates (glycolipids). And thus, these pose difficulty in the extraction of fat using simply grinding or drying techniques. In such cases, acid hydrolysis is commonly used, wherein sample is digested in presence of 3N HCl and heating conditions in order to release fat in easily extractable form.

Solvent extraction can be batch-type, semi-continuous or continuous, which are explained as follows:

1. Batch-type solvent extraction: In this method, sample is mixed with suitable solvent/solvent mixture in a separating funnel and mixed well to extract fat from aqueous phase/food into the solvent. Due to gravity or centrifugal effects, two phases gets separated (Fig. 11.16):

I. Lower, heavy aqueous phase

II. Upper, light solvent phase

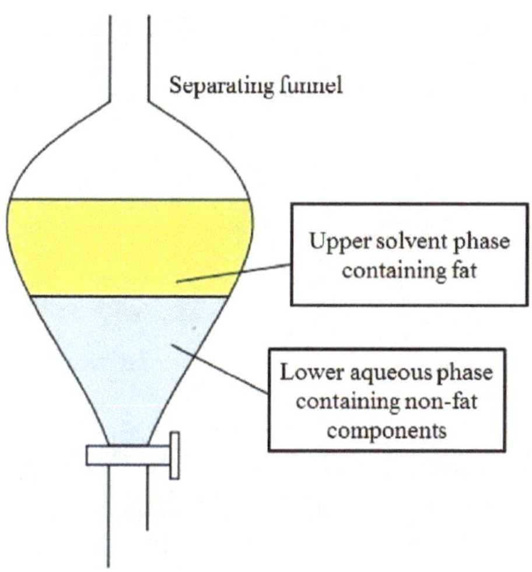

Fig. 11.16: A schematic diagram of separating funnel with two non-missible phases.

Upper phase contains extracted fat dissolved in solvent. Solvent is evaporated under low pressure/vacuum using hot plate and thus, extracted fat is recovered and calculated by using following formula:

$$\% \text{ fat (w/w)} = \frac{\text{Extracted fat (g)}}{\text{Weight of sample (g)}} \times 100$$

2. *Semi-continuous solvent extraction:* This method is used to enhance the extraction efficiency of organic solvents. Sox let extraction is the most common example of semi-continuous methods. In this method, ground sample is taken in a small and porous thimble (made of very thick or simple filter paper), which is placed in a vessel as shown below in the schematic diagram (Fig. 11.17):

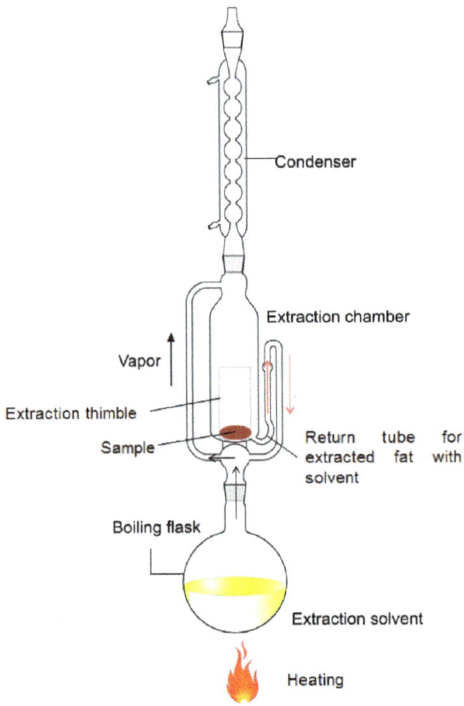

Fig. 11.17: Schematic diagram of the apparatus for fat extraction using Soxhlet method

Solvent is boiled and solvent vapors move toward the condenser, where they are condensed and come back downwards to the extraction chamber, where thimble is placed. Solvent, which is hot by temperature, drips into the thimble and starts the extraction of fat from the sample. After some time, when the extraction chamber gets filled with the condensed solvent, it returns to the boiling flask through siphoning process. This process continues for approximately 2-3 hours, and the fat is extracted from the sample into the solvent and collected in the boiling flask. After completion of process, boiling flask is removed from the apparatus and heated to evaporate solvent on a hot plate at lower temperatures (40-60°C). Thus, extracted fat remains in the flask as it is not evaporated at such lower temperatures. It is weighted and the amount is calculated using formula given above.

*3. Continuous solvent extraction:*Fat can be extracted continuously using Goldfish method, which is similar to the Soxhlet method, except that the extraction chamber is designed so that the solvent just trickles through the sample rather than building up around it. This reduces the amount of time required to carry out the extraction, but it has the disadvantage that channeling of the solvent can occur, i.e., the solvent may preferentially take certain routes through the sample and therefore the extraction is inefficient. This is not a problem in the Soxhlet method because the sample is always surrounded by solvent.

Non-solvent extraction methods

*1. Gerber method:*In this category, Gerber method is a common method which is used for the quantification of fat in milk and some milk products. First the sample is digested in presence of Gerber acid (90% sulphuric acid), which breaks the emulsion and facilitates the separation of fat from other components. Then isoamyl alcohol is used for the proper separation of fat and to dissolve it. The mixture of milk, sulphuric acid and isoamyl alcohol is centrifuged for some time and extracted fat in separated phase is measured. This method does not determine phospholipids as they remain in aqueous phase or at the boundary between the lipid and aqueous phases.

2. Babcock Method: The Babcock method is similar to Gerber method but does not involve the use of iso-amyl alcohol.

Supercritical Fluid (SCF) extraction

Supercritical fluid extraction technology is used to isolate heat-sensitive lipid components such as specific pigments, steroids, omega-3 and 6- fatty acids, etc. In this method, CO_2 is used as an extraction medium, which shows duel nature under pressurized conditions. When pressurized CO_2 is heated above a particular critical temperature, it becomes supercritical fluid and behaves both like gas as well as liquid. Being a gas, it can penetrate easily in food; and being a liquid, it can easily dissolve the fat/fatty components from the sample. In this method, food sample is heated under pressurized chamber and then mixed with supercritical CO_2 fluid. CO_2 extracts the lipid components and forms a separate layer from the aqueous phase. Now the temperature and pressure of the separated layer are reduced (released), which converts liquid CO_2into gas. Gas escapes leaving behind the fat components, which can be weighted and measured.

Nutritive value of milk

Milk is considered as an ideal food containing good quality protein, fats, sugar (lactose) and significant amount of vitamin and minerals. Except a few components such as iron and vitamin C, milk contains almost all the nutrients which are required for optimal growth by animals including humans.

1. ***Carbohydrates*:** Milk contains a unique sugar called lactose, which is not present in any other food except mammalian milk. It is a disaccharide, provides calories and plays a significant role in the synthesis of alpha-lactalbumin and other milk proteins.

2. ***Fats:*** Milk fat is an important source of essential fatty acids. Milk contains unique fatty acid: butyric acid, present only in milk fat. Butyric acid has been reported to show antioxidative and anticarcinogenic activities in animal studies. Milk fat also contains appreciable amount of fat soluble vitamins (A, D and E) and other biological active components such as conjugated linolenic acid (CLA). Milk fat contributes unique taste and flavor to different dairy products such as cheese, yoghurt, fermented dairy beverages, etc.

3. ***Proteins:*** Milk protein is considered high quality protein containing all essential amino acids. Milk protein shows good digestibility and high biological value as compared to plant proteins. Major milk proteins are casein and whey proteins. Major whey proteins are α-lactalbumin and â-lactoglobulin. Other whey proteins include serum albumin, immunoglobulins (IgA, IgG, IgM), protease peptones, lactoferrin and transferrin, which are reported to have biological activity. Nutritional composition of milk is shown below (Table 11.2):

Table 11.2: Nutritive value of milk (g/100g) of different species

	Energy (Kcals)	Moisture (g)	Protein (g)	Fat (g)	Mineral (g)	Fibre (g)	Carbohydrates (g)
Milk, buffalo	117	81	4	6	1	-	5
Milk, cows	67	87	3	4	1	-	4
Milk, goats	72	87	3	4	1	-	5
Milk, human	65	88	1	3	0	-	7

(*Source*: Gopalan *et al.*, 2004)

4. ***Micronutrients (Vitamins and minerals):*** Milk is a good source of water soluble vitamins particularly riboflavin and vitamin B_{12}. Riboflavin is important in carbohydrate metabolism and energy release reactions. It is also important for normal vision, immune functions and functioning of red blood cells. On the other hand, Vitamin B_{12} is essential for normal cell division and energy metabolism. In terms of minerals, milk is a significant source of calcium and phosphorus, which are crucial for normal bone and teeth formation. Calcium is also important for normal blood coagulation, energy yielding metabolism, proper

muscles contraction and nerve functions, digestive functions and regulation of cell division and differentiation.

Sample questions

1. Define calorie.
2. How will you measure the calorie content of any food using Bomb calorimeter?
3. Give a brief of Atwater system for measuring the energy value of any food.
4. What do you understand by quantitative and qualitative estimation of any food component?
5. Enlist the different methods for the quantitative estimation of carbohydrates and mention their specificity also?
6. Give the principle and chemical reaction involved for the following tests:
   a. Lane-Eynon method
   b. Anthrone method
   c. Phenol - Sulfuric Acid method
   d. Molisch's Test
   e. Fehling's Test
   f. Barfoed's Test
   g. Benedict's Test
   h. Seliwanoff's Test
   i. Bial's Test
   j. Iodine Test
7. Discuss the principle and procedure for following protein estimation methods:
   a. Kjeldahl method
   b. Biuret test
   c. Ninhydrin test
   d. Lowry's test
   e. Sakaguchi Test

8. Discuss various methods for lipid extraction and estimation methods.
9. Discuss Soxhlet extraction method for the estimation of lipids in foods.
10. What do you understand by Supercritical Fluid extraction? Where this method is used?
11. Give a brief of nutritive value of milk.

Suggested Readings

Gopalan. C., Rama Sastri B.V. and Balasubramanian, S.C. (2004). Nutritive Value of Indian Foods, National Institute of Nutrition, ICMR, Hyderabad.

Nielsen, S.S., 2017. *Food analysis laboratory manual*. Springer.

http://www.chem.boun.edu.tr/wp-content/uploads/2014/04/Chem-415-Experiment-1.pdf

http://ninindia.org/DietaryGuidelinesforNINwebsite.pdf

https://people.umass.edu/~mcclemen/581Proteins.html

http://people.umass.edu/~mcclemen/581Carbohydrates.html

http://people.umass.edu/~mcclemen/581Lipids.html

http://egyankosh.ac.in/bitstream/123456789/12041/1/Experiment-4.pdf

http://www.foragetesting.org/lab_procedure/sectionB/3/part3.1.htm

http://www.borrowfoundation.org/assets/uploads/THE%20NUTRITIONAL%20VALUE%20OF%20MILK%20BF.pdf

12

Digestion and Absorption of Carbohydrates, Lipids and Proteins

Introduction

Digestion may be defined as the hydrolytic process carried out by various enzymes for breaking down complex food components such as proteins, fats and carbohydrates into simpler molecules like amino acids, fatty acids and monosaccharides, respectively. Whereas, absorption may be defined as the movement of these simple soluble digested food components along with vitamins, minerals and water from gastrointestinal track into the circulatory system.

The process of digestion starts from oral cavity (mouth) and ends at large intestine including tongue, teeth, jaws, secretions from salivary glands (parotids, sub maxillary or submandibular and sublingual glands), pharynx, esophagus, stomach, small intestine and digestive enzymes secreted from pancreas and liver.

The oral cavity includes tongue, teeth, jaws and saliva. Tongue helps in ingestion and swallowing of the food while teeth help in mastication. The enzymes present in watery secretions of salivary glands (called saliva) facilitate the digestion of carbohydrates and mucin helps in lubrication of the food for swallowing. From oral cavity, the food goes to the stomach through a funnel shaped structure called pharynx and a tube called esophagus (Fig. 12.1). Here, the masticated food is mixed with hydrochloric acid and protein digesting enzymes. Then the food passes into the small intestine through pyloric region of the stomach. The small intestine is a highly coiled structure having three different regions viz. duodenum, jejunum and ileum. Most of the food components are digested and absorbed in the small intestine region of digestive track. The small intestine opens into a comparatively shorter and wider tube called as large intestine further differentiated into three parts viz. caecum, colon and rectum. The undigested part of the food passes from small intestine to the large intestine.

Most of the water and some B-complex vitamins are reabsorbed here from the undigested food and the remaining waste is excreted as fecal matter through anus.

Some Important Terms Associated with Digestion and Absorption

Fundus: Upper part of the stomach

Antrum: Lower part of the stomach

Gut: Inside of small and large intestine

Mucosa: Wet and slippery surface of an organ due to secretion of mucus

Epithelial membrane: A continuous layer of cells that covers the upper surface of an organ

Enterocytes: Most common cells of the epithelial membrane of gut

Goblet cells: Mucus secreting glands of the epithelial cells of the gut

Villi: Finger like projections about 0.025mm in height, present on the epithelial surface of the gut, giving a rough appearance.

Microvilli: Hundreds of very small projections present on the surface of enterocytes facing food, responsible for increasing the surface area for the purpose of maximum absorption of nutrients.

Brush border epithelium: Due to presence of villi and microvilli, the inner surface of gut looks like brush border and is known as brush border epithelium.

Ampulla of water: Pancreatic and bile duct joins together to form a common duct called as ampulla of water.

Gall bladder: Stores bile salts and releases them into duodenum as and when required.

Vagus nerve: Transmit signals from brain to stomach, pancreas and other digestive system components for getting prepared to start the digestion process.

Zymogens: Inactive form of enzymes

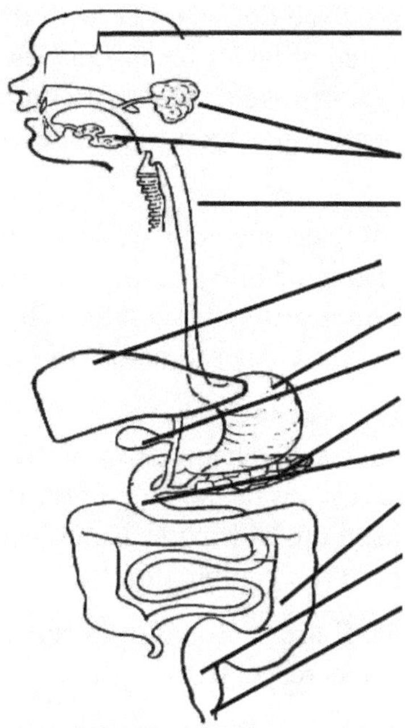

Fig. 12.1: Outline of Human Digestive Track

Gut hormones: These are mainly of three types:

(a) Gastrin, (b) Secretin and (c) Cholecystokinin (CCK)

Gastrin is a hormone with 34 amino acids, secreted from the G cells, situated in the antrum region of the stomach. Their role is to provoke the parietal cells of the stomach to release gastric acid. When gastric acid enters duodenum along with the digested food mixture, it stimulates the release of **secretin** hormone. Secretin provokes pancreas to secrete pancreatic juice and bicarbonate salts into the gut. Entry of dietary fat and proteins from stomach into small intestine stimulates **CCK's** release in blood stream. This leads to release of enzymes and zymogens from pancreatic juice into the gut.

Glands Secreting Gut Hormones

Salivary glands: They secrete saliva containing amylase and lingual lipase.

Gastric glands: These glands are located in the walls of the fundus. They contain parietal cells, chief cells, goblet cells and hormone secreting cells. Pepsinogen, mucus and hydrochloric acid (0.1N) are secreted by chief cells, goblet cells and parietal cells, respectively. In addition to HCl, an intrinsic factor is also secreted by parietal cells, which help in absorption of vitamin B_{12} from diet.

Pancreatic gland: Acinar cells are the dominant cells in the pancreas. They secrete various enzymes in zymogen or latent form. Besides acinar cells, duct cells and endocrine cells are also present in pancreatic gland. They are responsible for secreting bicarbonates, glucagon and insulin into the pancreatic juice, respectively.

Liver and Gall bladder: Hepatocytes, the cells of the liver synthesizes bile salts. Bile juice joins secretions from the bile duct of liver, the largest gland of our body. Bile juice contains bile pigments and bile salts. When there is no food, the bile juice is stored in a small muscular sac below the liver called as gall bladder.

Bile acids: Bile acids include cholic acid, chenodeoxycholic acid, deoxycholic acid and lithocholic acid. They are easily water soluble and are usually neutralized by adding sodium bicarbonate or sodium hydroxide. Neutralized acids are called as bile salts. They play an important role in lipid digestion.

Small intestine: Secretin and CCK are secreted by S cells and I cells of the duodenum part of the small intestine, respectively.

Absorption of the different nutrients from food occurs inside the small intestine via the luminal face of the mucosa. This face is covered by finger like projections called as villi. Each villus is further bordered by 8-10 small pouches or invaginations called crypts. The prime task of the cells of these villi is to absorb the dietary nutrients, left over of the secretions from various digestive glands and damaged and sloughed off cells. The villi are covered by cells called as enterocytes. These cells contains various membrane bound and intracellular digestive enzymes, various transport proteins for transporting the absorbed nutrients into the blood stream and enzymes used in triacylglycerol and chylomicron synthesis. The main purpose of the crypt cells is to secrete water for better dispersion & solubility and bicarbonates for neutralization of gastric juice. The cells at the epithelial surface of the small intestine have to divide and differentiate frequently so that its function is not hampered by the regular damage to epithelial cells from harsh environment of pancreatic secretions and enzymes. The crypt cells rapidly divide and moves upward, differentiates, develops microvilli and are eventually sloughed off. The whole cycle gets completed in 2 to 3 days period and takes place throughout life constantly. Other than this, there are also some endocrine and paneth cells, accounting for one percent of the total epithelial cells. Endocrine cells secret hormone whereas the paneth cells produce an antibacterial enzyme called as lysozyme. The composition of the plasma membrane of epithelial cells of the crypt and villi are different from each other and also from that of mitochondrial membrane and nuclear membrane in terms of cholesterol content. The plasma membrane of the villus cell contains a larger cholesterol/phospholipid ratio (0.85) than the plasma membrane of the crypt cell (0.60), making them more rigid.

Digestion and Absorption of Carbohydrates

Generally 45% of the total daily dietary energy requirements are fulfilled by carbohydrates. However, in countries where milk and meat products are not a part of regular diet, 70-90% of the total daily dietary energy requirements are fulfilled by them. Dietary carbohydrate includes monosaccharide (such as glucose, fructose and galactose), disaccharides (such as lactose, maltose and sucrose) and polysaccharides (such as starches and dietary fibres). Corn, potato, wheat, cassava root and rice are the major sources of starch in our diet. Starch is made up of small microscopic granules of further two components-amylose and amylopectin. In general, amylopectin constitute two third of the part. These granules are arranged in such an orderly pattern that digestion of raw foods having starch is impaired up to 50% by digestive enzymes. Cooking causes hydration, swelling and gelatinization of starch granules, making them more susceptible for enzymatic hydrolysis. Dietary fibre is that carbohydrate which is not completely digested in our digestive track as we lack the enzymes needed for their hydrolysis and digestion. This carbohydrate is acted upon by the microbiota present in the large intestine for their utilization.

***Digestion*:** All the carbohydrates present in the diet have to be hydrolyzed into the simplest monosaccharide unit called as glucose. Digestion of carbohydrates starts from the oral cavity. The enzyme α-amylase present in the salivastarts catalytic hydrolysis of starch in mouth. This enzyme is also secreted by pancreas through pancreatic juice in the form of pancreatic α-amylase. Both the enzymes viz. salivary and pancreatic amylase are α-1-4 glucosidases and assist in hydrolysis of only the internal 1-4 glucosidic bonds found in starch and glycogen. Moreover, the activity of salivary amylase is not as much significant as that of pancreatic amylase. From mouth the mixture of saliva & food (called as bolus) passes through esophagus into stomach. Here the food comes in contact with gastric juice. As hydrochloric acid secreted from the chief cells dominates in this juice, the pH of the food falls near to 2 and therefore, digestion of carbohydrates stops in stomach due to change in pH. The digestion is resumed in the duodenum where the food comes in contact with pancreatic juice. The α -1-4 glycosidic bonds present near the branching points and α-1-6 glycosidic bonds of amylopectin are not hydrolyzed by pancreatic amylase as explained earlier. Therefore, some intermediate digestion products are generated from the action of pancreatic amylase on starch such as dextrin, isomaltose and maltotriose. Dextrin is a hydrolysis product of starches containing upto 8 glucose units with one or more α 1-6 bonds. Isomaltose is a disaccharide with α-1-6 glycosidic linkage and maltotriose is a sugar molecule with three glucose units linked together by α-1-4 glycosidic linkage. The complete digestion of carbohydrates occurs in mucosal lining of the intestine by the action of α-dextrinase or isomaltase and sucrase. Both isomaltase and sucrase are present together bonded with one polypeptide chain. Isomaltase causes hydrolysis of α-1-6 glycosidic bonds. The isomaltase catalytic site is closest to N terminal of the chain, whereas sucrase site is towards C terminal. The other disaccharides like maltose, sucrose and lactose are acted upon by the enzymes maltase, sucrase and lactase, respectively and converted to monosaccharide units. The details of carbohydrate digestion have been summarized in the Figure 12.2.

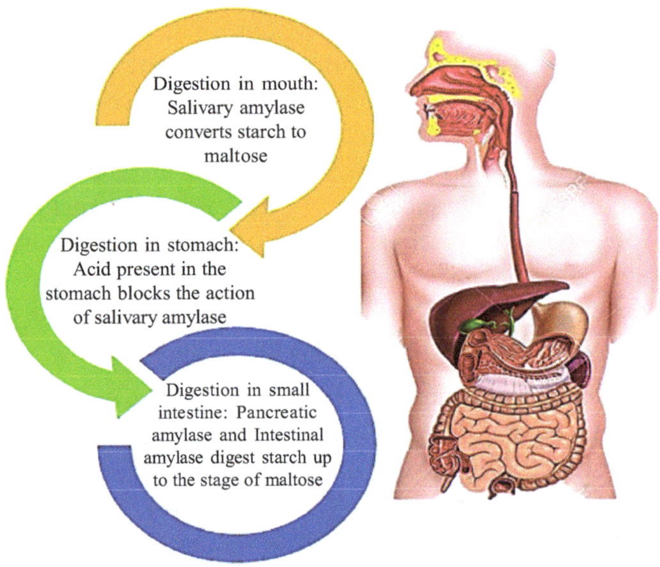

Fig. 12.2: Digestion of carbohydrates

Absorption: Glucose is the simplest carbohydrate that does not need further hydrolysis for absorption. Glucose gets absorbed from the intestinal lumen into the blood stream via a route through the mucosal epithelial cells. Glucose has to cross two membranes for reaching the blood stream. First membrane is the apical membrane and the second is basolateral membrane of enterocytes. Sodium ion dependent glucose transporter and GLUT-2 are the two transporter proteins needed by glucose molecules to cross these membranes, respectively. Sodium ion dependent glucose transporter is made up of 665 amino acids. Sodium ions act as the driving force in this transport mechanism. When there is a concentration gradient for sodium ions (the concentration of sodium ions is more towards outside than inside), the transporter protein facilitates transport of glucose molecules across the apical membrane of enterocytes. The same mechanism also operates in renal tubules of kidneys to prevent excessive loss of glucose through urine. GLUT-2 protein facilitates the transport of glucose across basolateral membrane of enterocytes towards the blood stream. This transport mechanism is an active one and requires energy in the form of ATP. Fructose is absorbed by a facilitated diffusion process using the transport protein called as GLUT-5. Other sugars are absorbed by simple diffusion through the lipid bilayer of the membrane (Fig. 12.3). In normal individuals, the digestion and absorption of carbohydrates is complete up to 95% or more.

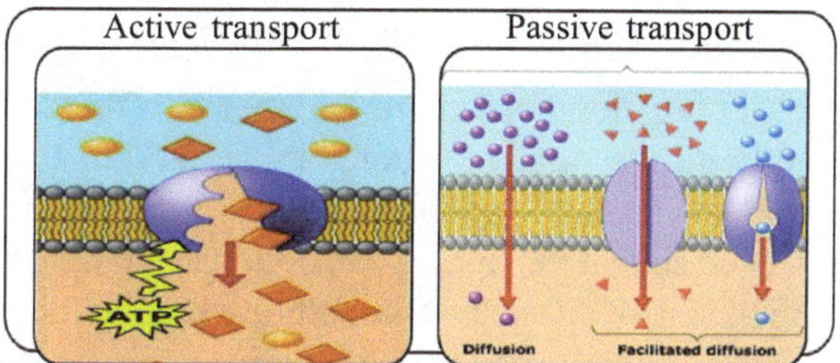

Glucose and galactose are absorbed into the mucosal cells by active transport, a process that requires energy and the involvement of a specific sodium dependent receptor The energy is supplied by ATP, and Na–is usually involved in the active transport mechanism. This is the fastest mode of transport

Dructose is transported into the mucosal cell by a specific facilitative transorter and can processed only down a concentration gradient. Inside the epithelial cell, most of the fractose is converted to glucose. The latter then enters the circulation.

Fig. 12.3: Absorption of carbohydrates

The main purpose of glucose is not just to cross the epithelial cell barrier and enter into the blood stream but its aim is to enter cells and get oxidized to provide energy. Our brain and central nervous system requires a steady supply of glucose for its proper functioning. GLUT-1 and GLUT-3 are the two transporter proteins helping in this process by allowing glucose to exit the blood capillaries of the brain and enter into the extracellular space so that the nerve cells can acquire their preferred energy food. The liver takes up glucose from the blood stream and converts it in to glycogen. During the period of starvation, liver converts this glycogen into glucose again to maintain the glucose level in blood. This transport of glucose from blood stream to liver and liver to blood stream is also regulated and facilitated by GLUT-2 transport proteins.

Digestion and Absorption of Lipids

Lipids are the nonpolar compounds exhibiting very low solubility in water. Generally, Lipids which are solid at room temperature are termed as fats and those which are liquid at room temperature are called as oils. Dietary lipids include triglycerides, phospholipids, cholesterol, cholesterol esters and fat soluble vitamins.

Triglycerides are made up of a backbone of glycerol alcohol and three fatty acids attached with it at its three ends via ester linkages. Depending upon whether all the three triglycerides attached are same or different, they are classified as simple and mixed, respectively.

As the lipids are the concentrated source for energy, the energy storage sites of plant and animal foods are the primary location for triglycerides, for example, seeds of the plants, adipocytes of the meat tissue etc. while the phospholipids are mainly located in the membranes that surrounds different cells and their organelles. Due to their bipolar nature, phospholipids play an important role in selective permeability of cell membranes for in and out movement of materials.

Digestion: Lipids as having non polar nature are not miscible in water and the enzymes needed for their digestion are water soluble, therefore, lipid digestion is somewhat different from digestion of carbohydrates and proteins. An emulsion of fat and water is needed so that the lipases can better act on lipids. Stabilization of this emulsion is achieved by bile salts. Bile salts are bipolar molecules, having both hydrophobic and hydrophilic sites. The hydrophobic sites gets attached with the lipid phase while the hydrophilic sites gets in contact with the aqueous phase, thereby preventing aggregation of the fat droplets or micelles resulting in stability of the emulsion.

The lipases responsible of lipid digestion are secreted from stomach and pancreas as gastric and pancreatic lipases. In case of adults, the major digestion is carried out by pancreatic lipases. However, during infancy, the secretion of pancreatic juices is not that much and therefore, the responsibility of digesting dietary lipids are mainly of gastric lipases. This is also the reason why low fat diets are advised for infants by dietitians.

Digestion of lipids does not start from oral cavity as our saliva does not contain any lipases. The digestion of these dietary lipids happens very little (10-30% of triglycerides) even in the stomach. Although some gastric lipases are present in the stomach but they are inhibited by absence of bile salts and the acidic secretions of the stomach. However, the peristaltic waves generated by forceful contractions of the stomach break up these dietary lipids into small droplets or micelles. These micelles then enters small intestine, where they gets solubilize and emulsified with bile salts secreted in pancreatic juice along with pancreatic zymogens, enzymes and a small molecular weight protein called as colipase. Due to this emulsification, the fat droplets further break up, resulting in increase in the ratio of their surface area to size several fold. The detergents of bile salts also tend to form a layer around triglycerides reducing their accessibility to pancreatic lipase. Colipase, secreted in almost equal amount to pancreatic lipase, helps in anchoring the pancreatic lipase to triglyceride molecules. Now, the lipase enzymes from the secretions of the pancreatic juice act on these emulsified fat droplets. The triglyceride molecules dispersed as micelles gets hydrolyze into 2-monoglycerides by the action of pancreatic lipase. The third fatty acid attached with glycerol backbone is cleaved by esterase but this is not an essential step for absorption. Then the bile salts direct these 2-acylglycerols and free fatty acids towards absorptive surface of the gut mucosa. Besides this, the other lipids like cholesterol esters and short chain triacylglycerol are hydrolyzed by cholesterol esters and other less specific lipid esterase, respectively. Similarly, Phospholipase A_2, secreted as proenzyme and activated by enzyme trypsin causes hydrolysis of phospholipids.

The bile salts play an important role is lipid digestion. The total amount of bile salts produced in our body is 2-4g. However, the amount needed on daily basis is 20-30g. The ten times more amount in demand is met by recycling of bile salts from intestine to liver and this process is called as enterohepatic circulation of bile salts. Actually some of the bile salts from intestine diffuse back into the food particles to facilitate further digestion and the rest are transported down to ileum, where they get reabsorbed back into the portal vein of the liver through blood stream. Sodium ions and bile acid co-transporters help in their absorption from ileum. Difference in the sodium ion concentration in ileum and enterocytes, directs them towards the apical membrane of the enterocytes, from where they gets entry into enterocytes. The bile salts then cross the basal membrane barrier of enterocytes and enter into the blood stream. Finally, liver guides them from blood stream to gall bladder for temporary storage until the duodenum receives food for digestion.

Absorption: The free fatty acids and 2-mono glycerides produced as a result of enzymatic action enters the enterocytes by crossing the plasma membrane. Here, the free fatty acids gets activated by attaching with Coenzyme-A. These activated fatty acids get attached to the Sn-1 and Sn-3 positions of mono glyceride to form triglycerides again. Now these reassembled triglycerides and other digested lipid components are transported from enterocytes into the blood stream through special carrier proteins called as Apo lipoproteins. These proteins are mainly of two types- Apolipoprotein A and Apolipoprotein B. These proteins combine with the reassembled triglyceride molecules and the complex formed is called as chylomicron. The core of these chylomicrons consists of triglycerides with a protein layer surrounding them. These chylomicrons get incorporated into the secretory vesicles. The secretory vesicles containing the chylomicrons move across the cytoplasm towards the basal membrane of the cells. Then these vesicles fuse with the basal membrane and their content is poured out into the intercellular space. The secreted chylomicrons diffuse through pores, channels and intercellular space to enter into a structure called as lacteal. Lacteal is a capillary like structure having two ends. One end touches the tip of the villus and the other one is connected with a very fine network of branching tubules, ultimately leading towards the thoracic duct. The chylomicrons are pushed slowly through the lymphatic system up to the thoracic duct. From thoracic duct, the chylomicrons are transported into the blood stream through left subclavian vein, located near the left shoulder. These triglycerides along with other digested lipids, gets packaged into particles called chylomicrons. This packing allows them to enter in to an extracellular fluid called as lymph.

Digestion and Absorption of Proteins

Protein, a large and complex biomolecule undergo a series of reactions and is acted upon by hydrochloric acid, proteases secreted by gastric and pancreatic juice to got separated into free amino acids. From nutrition point of view, it's the costliest and important part of the diet as it contains amino acids needed for repair, growth and maintenance of our body tissues. As discussed earlier, few amino acids are essential as they cannot be synthesized by us. The diet should contain all these amino acids. Therefore, along with quantity, quality of diet also matters a lot.

Digestion: The dietary proteins get homogenized by the jaw mastication. These homogenized proteins are then denatured by the gastric acid. Denatured proteins are easily hydrolyzed by the pepsin. However, cooking also denatures proteins in the diet and helps in easy digestion. Along with protein denaturation, gastric acid also plays an important function of activating the enzyme pepsin. Pepsin causes partial hydrolysis of the protein molecules by breaking the peptide bonds in between the protein chains producing peptide fractions. The partially hydrolyzed proteins and peptides are moved towards duodenum. Their appearance in duodenum stimulates release of cholecystokinin into the blood stream, provoking pancreas to release different zymogens such as trypsinogen, chymotrypsinogen, proelastase and procarboxypeptidase into the pancreatic juice. Enterokinase activates these zymogens by shading a small part of their chain into their respective active enzymes such as trypsin, chymotrypsin, elastase and carboxypeptidase. The dietary proteins are further hydrolyzed by the activated proteolytic enzymes into free amino acids and oligopeptide fragments containing 2 to 8 amino acids. Details of protein digestion have been summarized in Fig. 4.

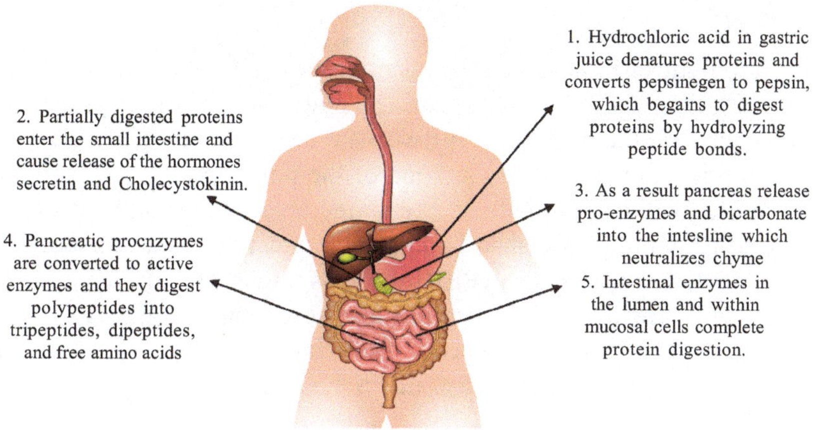

Fig. 12.4: Digestion of proteins

Different vitamins and minerals, if bound with the dietary proteins also getseparated by the action of gastric acid and proteolytic enzymes. Each proteolytic enzyme released from pancreatic juice is having different specificities based on a specific bond or nature of side chain. For example, trypsin is specific for the carboxylate group belonging to basic amino acids such as lysine and arginine. Chymotrypsin's specificity resembles that of pepsin. It prefers to act on the carboxylic group donated by aromatic amino acids. Elastase is little bit broad in its specificity and prefers to attack bonds next to small amino acids such as serine, valine, phenylalanine and glycine. Carboxypeptidase comes under the category of exo-peptidases and prefers to cleave the peptide linkage starting from carboxyl end of the chain. The origin of different proteolytic enzymes from different parts of digestive system is as shown in Fig 12.5.

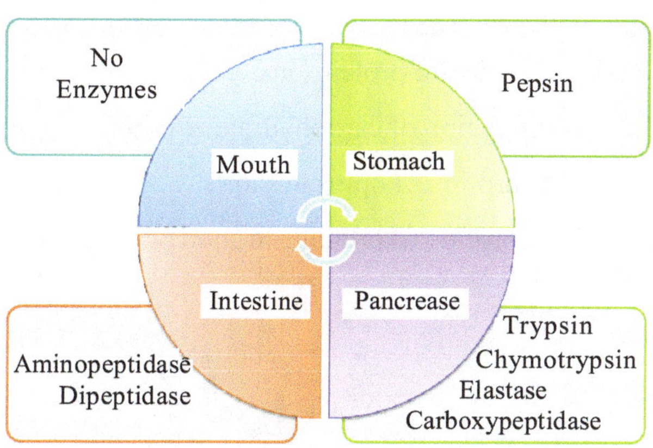

Fig. 12.5: Enzymes involved in protein digestion

Absorption

Some proteolytic enzymes like aminopeptidase are also attached with the brush border epithelium of enterocytes, causing hydrolysis of the peptides coming with contact with them. Aminopeptidase are also exopeptidases cleaving the peptide bonds from the free amino end of the polypeptides. The products of dietary proteins digestion include free amino acids, di, tri and tetra peptides. The di and tri peptides are absorbed through intestine via a transport protein called as peptide transporter 1 (PEPT-1). This transport is driven by influx of hydrogen ions. Besides this, there are also four other transport systems for carrying free amino acids into blood via sodium ion influx. These four sodium ion specific transport systems are as mentioned below:

A. Acidic amino acids such as glutamate and aspartate

B. Basic amino acids such as lysine and histidine

C. Neutral amino acids such as alanine, glycine and serine

D. Glycine, proline and hydroxyproline

One more protein transport molecule called as PEPT-2 is also present in renal tubules, responsible for reabsorption of di and tri peptides from urine back into the blood stream.

Sample questions

1. Define the terms digestion and absorption.
2. Discuss briefly gut hormones and the glands secreting these hormones.
3. Discuss about digestion of carbohydrates in detail.
4. Discuss about digestion of lipids in detail.
5. Discuss about digestion of proteins in detail.
6. Discuss about absorption of carbohydrates in short.
7. Discuss about absorption of lipids in short.
8. Discuss about absorption of proteins in short.

Suggested Readings

Adler-Nissen, J. (1986). Enzymic Hydrolysis of Food Proteins. Elsevier applied science publishers.

Akoh, C. C. (Ed.). (2017). Food Lipids: Chemistry, Nutrition, and Biotechnology. CRC press.

Boland, M., Golding, M., & Singh, H. (Eds.). (2014). Food Structures, Digestion and Health. Academic Press.

Cox, M. M. and Nelson, D. L. (2008). Lehninger Principles of Biochemistry. 5th edn. New York: Worth Publishers.

Damodaran, S. (Ed.). (2013). Food Proteins and Lipids (Vol. 415). Springer Science & Business Media.

Hudson, B. J. (2013). Biochemistry of Food Proteins. Springer Science & Business Media.

Regenstein, J. (2012). Food Protein Chemistry: an Introduction for Food Scientists. Elsevier.

Satyanarayana, U. and Chakrapani, U. (2013). Biochemistry. 4th ed. Kolkata: Elsevier Books and Allied (P) Ltd.

Sullivan, R. J. (2009). Digestion and Nutrition. Infobase Publishing.

Talwar, G. P., Srivastava, L. M. and Moudgil, K. D. (1989). Textbook of Biochemistry and Human Biology, 2nd edn. New Delhi: Prentice Hall of India Pvt. Ltd.

https://www.diet.com/g/digestion-and-absorption

https://www.slideshare.net/enamifat/introduction-to-digestion-and-absorption-local-hormones-of-git-different-digestive-juices

http://rajkumarbiology.weebly.com/class-11-chap-16.html

13

Milk Intolerance and Hypersensitivity

Introduction

There are many people all over the world who have a certain level of sensitivity or intolerance towards milk and milk products. The reactions to milk which involve the immune system are termed as milk allergy, whereas, difficulty in the digestion of the milk and milk products is known as milk intolerance. Milk intolerance occurs due to malabsorption of milk sugar (lactose) in intestine, a condition in which lactose is not digested into its monomers (galactose and glucose) due to lower activity or insufficiency of lactase enzyme. Inability to digest lactose, which is a disaccharide present only in milk, is specifically known as lactose intolerance. However, hypersensitivity involves the adverse immune reaction (allergic reaction) in the body caused due to the ingestion of milk proteins. Milk and milk products are major source of basic as well as essential nutrients and cannot be skipped from the diet of the individual. It is a rich source of calcium, which is essential for the growth and reformation of bones. The decline in calcium may lead to thin, fragile bones that break easily, a condition called osteomalacia or osteoporosis in women. Thus, it is a matter of concern for both children and adults who are allergic or intolerant towards milk and milk products and not getting enough calcium and other nutrients in diet. This chapter would provide more insight about milk intolerance and milk allergy/hypersensitivity and how to deal with it.

Lactose Intolerance

Lactose is a milk carbohydrate found in all types of milk including human. Under normal conditions, lactose is converted into galactose and glucose in small intestine by the action of lactase followed by the transport to the blood through intestinal mucosa. Lactose intolerance is the condition in which a person is unable to digest milk sugar (lactose) properly due to deficiency of lactase (β-galactosidase). In such a condition, lactose remains undigested and reaches to the colon, where it is degraded into small acids such as butyric acid, acetic acid, propionic acid and produce H_2 and other gases due to fermentation by local microflora. The other symptoms include vomiting, diarrhea, bloating, gas, nausea and pain in abdomen.

In the world, around 75% population suffers from lactose intolerance. The enzyme activity of lactase remains high in neonates and declines with the age. The activity of lactase might be low enough naturally, which is strongly related to ethnicity, with highest rates of lactose malabsorption in Asian populations, Native Americans and African Americans (60–100%) and lowest rates in people of northern European origin and the US white population (2–22%). Sometimes, an allergy or infection in small intestine may cause reduced secretion of β-galactosidase, resulting partial digestion of milk and milk products. If this is the condition, a person may take weeks or months for full recovery. Children having the lactose intolerance, naturally begin to produce lesser amount of β-galactosidase in later years. Some children show pronounced decline in enzyme production or terminate totally. The clinical signs of lactose intolerance are often expressed in adolescence or early adulthood. The enzyme β-galactosidase is originated by the cells that line the small intestine i.e., duodenum. In people, who lack β-galactosidase, usually start to develop the symptoms associated and said to be lactose intolerant. Its symptoms usually start to show-off after few hours or days of consuming lactose containing foods or drink.

Mechanism of Lactose Intolerance

Normally the hydrolyzed products of lactose such as glucose and galactose are taken up by the mucosa of the small intestine. But, in lactose intolerant people, depending upon the severity of condition or amount of β-galactosidase activity, certain mass of lactose would not be hydrolyzed (digested) by β-galactosidase in to simpler forms such as glucose and galactose and this lactose reach the colon intact; where fermented by intestinal microflora and results in production of short chain fatty acids viz. butyric acid, propionic acid, acetic acid and gases viz. CO_2, H_2 and methane. The undigested lactose increases the osmotic pressure as well as lactic acid produced is also osmotically active and draws water into the intestines resulting in diarrhea. The osmotically active constituents in intestine draw more water. The manifestations like a headache, loss of concentration, amnesia, muscle and joint pain, severe weakness and fatigue, allergies, cardiac arrhythmia, ulcers, sore throat and increased frequency of urination occur. In the presence of these signs, it is essential to analyze that these are the result of lactose intolerance or coincidence or are due to some other reasons.

Types of lactase (β-galactosidase) deficiency

There are four types of lactase deficiency:

a) ***Primary hypolactasia***: This one is the most common type of lactase deficiency and is genetic, upsetting mostly adults. The persons suffering from this condition have lesser production of β-galactosidase in the body. This downturn often begins at about age of two or may begin in later ages. This is also known as primary lactase deficiency.

b) ***Secondary hypolactasia:*** It mainly damages the small intestine through infection, an autoimmune disorder, inflammatory bowel disease or other diseases as well as by other environmental causes, which may affect the enzyme secretions in small intestine. Curing the intrinsic cause normally improves the condition, thus, reversible. It may occur in both infants and lactase persistent adults. This type is also known as acquired hypolactasia.

c) ***Primary Congenital alactasia:*** It is also known as congenital lactase deficiency. It is an exceptionally rare genetic disorder in which either very little enzyme is present or enzyme is entirely absent from the small intestine since birth. People with this type of deficiency would not be able to assimilate lactose from birth, even breast milk. This genetic defect is typified by a complete deficiency of β-galactosidase enzyme.

d) ***Developmental lactase deficiency:*** It usually occurs in premature babies and remains for only a short time after birth.

Diagnosis of Lactose Intolerance

The below-given test determines the presence of lactose intolerance:

a) ***Lactose Tolerance Test:*** This usually requires fasting before the test and then drinking a lactose solution. Blood sugar level is measured by taking several blood samples over a two hour period. In normal condition, when lactose reaches the digestive system, β-galactosidase breaks it into simpler forms such as glucose and galactose. In liver, galactose changes into glucose which then enters into the bloodstream and the person's blood glucose level increases. However, in lactose intolerance condition, where it is incompletely digested, the blood glucose level is not raised and diagnosis is confirmed.

b) ***Hydrogen Breath Test:*** This test work by determination of the amount of hydrogen in the breath of a person. In a normal individual, very less amount of hydrogen is measured. However, in lactose intolerant person, the undigested lactose reaches the colon intact and fermented by microorganisms present and produce several gases including hydrogen. The hydrogen produced by the intestines is absorbed into the bloodstream and reaches the lungs, and then exhaled. During the diagnosis, a person is allowed to drink lactose-containing beverage and breath is analyzed at a regular interval. Increased levels of hydrogen in the breath signal incomplete digestion of lactose. As lactose intolerance causes diarrhea in infants which may lead to dehydration and may be fatal, so it could be dangerous for newborn babies. Therefore, lactose tolerance and hydrogen breath test are not performed on newborns.

c) ***Stool Acidity Test:*** This test usually determines the stool acidity and can be used for both infants and young children. When undigested lactose arrives in the colon and gets fermented by bacteria; it leads to the production of various acids such as lactic acid and some fatty acids. Thus, measuring the acidity of stool sample would help in determining the condition.

d) ***Blood Test:*** In this test, person's blood glucose level is measured routinely after ingesting lactose-rich drink. A typical elevation of 50% to 100% in blood glucose level within one to two hours in lactase persistent individuals while there is no rise in blood glucose level in lactose intolerant individuals. However, this test requires the frequent blood sampling, so this approach has not been used much.

e) ***Intestinal Biopsy:*** This test is usually done after confirmation of positive result in hydrogen breath test. In this test, lactase enzyme presence is identified by upper gastrointestinal endoscopy.

f) ***Stool Sugar Chromatography:*** The chromatography technique is used to isolate and identify undigested lactose present in person's fecal sample. However, this test is not considered reliable enough for conclusive results.

g) ***Genetic Diagnostic:*** This test may be useful in assessing primary hypolactasia. In lactose intolerance condition, a person has two polymorphisms located in the *MCM6* gene, C/T 13910 and G/A 22018.The detection of the polymorphisms is done by molecular biology techniques from extracted DNA. This test determines a non-invasive definitive diagnostic.

Management of Lactose Intolerance

a) ***Adding β-galactosidase externally:*** Addition of β-galactosidase enzyme externally in milk and milk products may help in combating such condition. The enzymes are available in both tablet and liquid form and can be used when required.

b) ***Lactose-free and lactose-reduced milk and milk products:*** These products can be found at global markets and are nutritionally similar to regular dairy products. The manufacturers treat milk with the β-galactosidase enzyme and make it lactose-free by hydrolyzing it in simpler forms such as glucose and galactose; thereby reducing the chances of developing digestive problems. The lactose-free milk has a shelf life similar to normal milk or if it is ultra-pasteurized, even longer than normal milk. However, it may have a slightly sweet taste as compared to the normal milk.

c) ***Milk products with little or no lactose:*** Yogurt is manufactured by bacterial fermentation using the culture of *Lactobacillus delbrueckii subsp. bulgaricus* and *Streptococcus thermophiles*; and thus, it contains little or no lactose as it is converted into lactic acid and other acids. Lactose intolerant individuals may tolerate yogurt well. In addition, it also contains the enzyme β-galactosidase produced by the bacterial cultures which may help in better digestion of remaining lactose in the intestine. Other than yogurt, certain other fermented milk products like kefir, koumiss, leben, labaneh, acidophilus milk, mursik, and viili and *dahi* etc. could also be used as a substitute of milk.

Milk allergy

Milk protein allergy is often confused with milk intolerance since both are caused by the same dietary source milk; as well as similar in symptoms such as diarrhea, abdominal pain or cramping. However, lactose intolerance and milk allergy are not related. Lactose intolerance is not an allergic reaction; it is just sensitivity to milk and milk products due to deficiency of β-galactosidase, which has been discussed earlier. However, milk allergy is triggered by the immune system and an allergic reaction to the milk proteins ingested through milk. Given this, it is necessary to differentiate between these two situations. Depending upon the condition of intolerance, a person can either limit the consumption of lactose-containing foods or totally avoids; while in milk allergy, any food containing milk proteins must be excluded from the diet.

Cow's milk protein allergy (CMPA) is an adverse immune response to the antigen present in cow's milk and commonest type of food allergy primarily affecting children, especially infants. In this condition, cow milk proteins or fractions like casein, α-lactalbumin, and β-lactoglobulin are not recognized by the body and are considered as allergens. The increased use of cow's milk instead of human milk for infants' feeding has resulted increased number of incidences related to this disease. In milk proteins, about 80% part is casein fraction which might acts as a potent allergen, causing greater sensitivity in individuals. Casein is further classified into αs1, αs2, β, and κ-casein. When casein is digested, it contains proteins that are transformed into β-casomorphins, an opioid compound. A1 allele of β-casein binds with the β-casomorphins and it is believed that its ingestion is the cause of the allergic reaction in human. In contrast, an A2 allele of β-casein has no role in causing an allergic reaction. Some breeds of cow have a higher expression of an A2 allele of β-casein and hence produces less allergenic milk. Thus, in recent time, several companies are selling A2 milk, offering health promoting attributes with decreased health problems. Although our body has the ability to assimilate milk proteins; sometimes, these proteins are treated as antigen; thereby stimulates the allergic response. At birth, enteric enzyme system of digestive tract plus the immune system are not fully developed and still maturing. At this time, the most ideal and convenient food for infants is breast milk. The small part of proteins and peptides consumed by the lactating woman are absorbed into the milk; thereby the newborn slowly develops the immune system by remaining in the acquaintance with the food; thus helps to assimilate the future foods they will consume.

As per the mechanism of the allergic response, food allergies are divided into three groups:

a) IgE-mediated: This is the most common type of food allergy results in acute occurrence of the symptoms shortly after ingestion and may advance into anaphylaxis (severe allergic reaction) that need to be treated as soon as possible. Symptoms may include rash, hives, itching on mouth area, and swelling in lips, tongue, eyelids, whole face, difficulty in swallowing or congestion, runny nose, hoarse voice, wheezing, shortness of breath, abdominal pain, diarrhea, dizziness, fainting, nausea, and vomiting. These adverse allergic reactions differ from case to case or individual to individual. The severity is seen when allergies start affecting the respiratory tract or blood circulation. Anaphylaxis is marked by the manifestations such as wheezing, congestion, cyanosis, weak pulse, pale skin, and fainting. These symptoms are shown when primarily IgE antibodies are involved. When untreated, this can further lead to anaphylactic shock due to vasodilation (a low blood pressure condition) or sometimes death which is very rare.

b) Non-IgE mediated: This is characterized by an immune response not implying immunoglobulin E (IgE) and may occur after hours or days of assimilating food (containing milk proteins); thus, complicating the diagnosis of the condition and mislead the treatment process.

c) IgE and non-IgE-mediated: This is a hybrid of the two types mentioned above.

The prime allergen in casein are phosphorylation sites, thus, the allergenicity could be attributed to changes in these regions. In contrast to whey protein, caseins are found to be predominantly allergenic in people with IgE-mediated cow milk protein allergy. β-casein contains 209 amino acids and has thirteen genetic variants or polymorphisms such as A1, A2, A3, B, C, D, E, F, H1, H2, I and G. The allele variants A1 and A2 of β-casein are the most common in dairy cows.

A1 and A2 variants are distinguished by the change of one nucleotide at the position 67 in the chain (A1 has histidine amino acid and A2 has proline amino acid at 67 positions). Studies have confirmed that initially, the A2 allele is the dominant one in entire bovine population, and A1 allele was the result of mutation. Due to these subtle differences in the structures of β-casein variants, they are digested differently. A2 allele of β-casein hydrolyzed at a very low rate or not hydrolyzed at all by the enzyme, leading to the formation of a peptide, β-casomorphin-9 (BCM-9) while, the digestion of A1 allele of β-casein produces the exogenous opioid peptide, β-casomorphin-7 (BCM-7). In humans, BCM-7 is thought to be a primary cause for health-related issues. Therefore, the intake of A1 milk has been linked with a higher incidence of diseases such as diabetes mellitus type-I, coronary heart disease, arteriosclerosis, schizophrenia, autism and sudden infant death syndrome as well as allergies. In contrast, studies confirm that the A2 milk does not cause any health implications or problems.

As discussed above, milk protein allergy can occur by both IgE-mediated and non–IgE-mediated pathways. An IgE-mediated allergy occurs when IgE antibodies are accompanied with mast cells (a type of white blood cell) and bind the antigen. Histamine, a potent inflammatory mediator is released from the mast cells when the IgE antibodies bind with an antigen resulting in an immediate allergic response by the body. On the other hand, Non–IgE-mediated milk protein allergy causes direct stimulation of T cells by milk protein antigens and antibodies included are IgA or IgG. These more complex immune interactions result in the release of cytokine which elevates the production of antibodies that recognize the milk proteins as an antigen by signaling inflammatory response. The distinction between IgE-mediated and Non-IgE mediated mechanisms is necessary, as IgE-mediated milk protein allergy is associated with multiple food allergy and is of higher risk, as well as atopic conditions such as asthma later in life.

The milk allergy can be prevented by avoiding milk proteins while maintaining a balanced and nutritious diet. Studies have shown that other protein sources can be included in order to complete diet. Other treatments like fermentation or heating can reduce allergenic potential of proteins by converting it into small peptides, amino acids or denatured proteins, which are less likely to cause an allergenic reaction.

Phenylketonuria

Phenylketonuria (PKU) is a disorder showing the inability to metabolize phenylalanine due to deficiency of phenylalanine hydroxylase (PAH) enzyme. Phenylalanine, an essential amino acid, is converted into tyrosine under physiological conditions. Due to deficiency of PAH enzyme, phenylalanine is not converted into tyrosine leading to the excessive accumulation of phenylalanine in blood, urine and other tissues as well as in the brain. In addition to phenylalanine, some derivatives like phenyl ketones also accumulate in the body. As phenyl ketones can be excreted through urine, thus the name, Phenyl-Ketone-Uria (PKU) or phenylketonuria.

Aspartame, an artificial sweetener, is a non-food source of phenylalanine. When metabolized in the body, this compound forms several chemical compounds including phenylalanine. Thus, all the food products that subsume aspartame should be labeled with a warning: "Phenylketonurics: contains phenylalanine or "Contains a source of phenylalanine". Thus, labeling the food products help people to avoid such foods. Additionally, Tyrosine, a non-essential amino acid, is a precursor for neurotransmitters and hormones. It is also required by newborns for the synthesis of proteins and neurotransmitters. PKU can be diagnosed by testing the amino acids in the blood sample and PAH gene mutations. Some countries have a screening program for detection of phenylalanine in infants, thus can be treated by showing symptoms of the disease. The treatment includes restricting the food containing phenylalanine such as proteins from milk, meat, fish, eggs and other types of food that contain phenylalanine or substituting these proteins with a diet that contains all the other amino acids except phenylalanine. Additionally, the diet can be complemented with tyrosine in order to avoid its deficiency.

If untreated, a person would have a musty or mouse-like odor due to the extensive accumulation of phenylalanine. The excessive amount of phenylalanine may cause brain damage or other serious complications, as it affects brain nerves. The severe form of the disease is classic PKU, which occurs when PAH activity is completely reduced or absent. The milder form of this disease is non-PKU. People with a very mild form of PKU may not require any treatment with a diet containing low phenylalanine content.

Galactosemia

Galactosemia is a rare disorder resulting from an enzyme deficiency that is responsible for the metabolism of galactose. The main dietary source of galactose is lactose, a carbohydrate found in milk. When lactose is metabolized, it breaks down into glucose and galactose. In absence of specific enzyme, galactose cannot be broken down and it starts accumulating to toxic levels in the body. Such condition is known as galactosemia (meaning too much galactose in the blood). The usual symptoms are diarrhea, dehydration, jaundice, liver failure, hypoglycemia, cataracts, slow development and ultimately death. Among infants having galactosemia, the sepsis due to *E. coli* seems to be particularly frequent and is the usual cause of death. The best way to treat the disease is the withdrawal of all foods from the diet that contains lactose and/or galactose.

There are several variants of this disorder such as:

a) ***Type 1:*** Also known as 'Classic Galactosemia', is a rare genetic disorder due to an extreme deficiency of the enzyme Galactose-1-Phosphate Uridyltransferase (GALT).

b) ***Type 2:*** It is caused by the deficiency of galactokinase, a rare enzymatic defect. This may result in cataracts during infancy and probably mild mental retardation.

c) ***Type 3:*** It is caused by deficiency of UDP galactose epimerase and UDP-galactose-4-epimerase resulting in deposition of galactose-1-phosphate in the red blood cells and is normally benign.

Galactose is metabolized by 3-enzyme Leloir pathway. In absence of specific enzymes mentioned above, galactose starts accumulating and is reduced to sugar alcohol (galactitol). However, galactitol is not an appropriate substrate for other metabolic enzymes; thus, starts to accumulate in body tissues and is excreted in the urine of galactosemic patients. The piling up of galactitol in eye lens has several adverse effects such as cataract or classic galactosemia.

The only possible therapy for classic galactosemia is the withdrawal of lactose and galactose from the diet. This is usually accomplished by switching the baby from drinking breast milk or a milk-based formula to drinking a low galactose formula, such as soy or elemental formula. Even with an early diagnosis and a restricted diet, some people still experience long-term complications such as learning disabilities, speech difficulties, neurological impairment as well as ovarian failure due to galactosemia.

Sample questions

1. What do you mean by lactose intolerance? How it differs from milk allergy?
2. Explain lactose intolerance, its effects and treatment in detail.
3. Briefly explain the different tests to measure lactose intolerance.

4. What do you understand by PKU? What are the measures taken to combat this disease?
5. How PKU derived its name?
6. What is galactosemia? Explain different types of galactosemia disorders.
7. What are the symptoms of galactosemia? What measures are taken to treat the galactosemia?
8. Define milk allergy?
9. Differentiate between milk allergy and lactose intolerance.
10. What are constituents of milk responsible for milk allergy?
11. Differentiate between A1 and A2 milk?
12. What is the mechanism for the occurrence of lactose intolerance?
13. How can lactose intolerance be managed through diet?

Suggested Readings

Walstra, P., Wouters, J. T. M. and Geurts, T. J. (2006). *Dairy Science and Technology*. 2ndedn. Parkway NW: CRC Press (Taylor & Francis Group), pp. 551-573.

Abumrad, N. N., Schneider, A. J., Steel, D. and Rogers, L. S. (1981). Amino acid intolerance during prolonged total parenteral nutritional reversed by molybdate therapy. *Am J Clin Nut*, 34, pp. 2252.

Eskin, N. M. A. and Shahidi, F. (2013). *Biochemistry of foods*. 3rdedn. USA: Academic Press Elsevier.

Fox, P. F. and McSweeney, P. L. H. (1998). *Dairy Chemistry and Biochemistry*. New York: Blackie Academic and professional (Chapman and Hall).

Fox, P. F. and McSweeney, P. L. H. (2003). *Advanced Dairy Chemistry-1. Proteins*. 3rdedn. New York: Kluwer Academic.

Fox, P. F. and McSweeney, P. L. H. eds. (2006). *Advanced Dairy Chemistry-2. Lipids*. 3rdedn. New York: Kluwer Academic.

Fox, P. F. and McSweeney, P. L. H. eds. (2009). *Advanced Dairy Chemistry-3. Lactose, Water and Minor Constituents*. 3rdedn. New York: Kluwer Academic.

Sardesai, V. (2012). *Introduction to Clinical Nutrition*. 3rd ed. New York: CRC Press.

14

Planning and Nutritional Policies

Introduction

Spectacular accomplishment in every sector has enabled India to be among top achieving nations. However, every section of the society is not benefited by these achievements and problem of malnutrition still remains unresolved. Malnutrition is a common problem faced by underdeveloped and developing countries such as Philippines, Nepal, Sudan, Tanzania and India since long back. Malnutrition includes undernutrition (wasting, stunting, underweight), inadequate vitamins or minerals as well as overweight, obesity and resulting diet-related non communicable diseases (NCDs). As per the latest report of World Health Organization, 52 million children under 5 years of age are wasted, 17 million are severely wasted and 155 million are stunted, while 41 million are overweight or obese (WHO, 2018). More than 50% women of marriage age are suffering from anemia. If women are malnourished at the time of pregnancy and delivery, the baby born face the problems such as physical retardation, sensitive to common infections, reduced cognitive abilities, reduced mental development, etc. As per the World Health Organization (WHO) guidelines, starting from the date of conceiving a child, the first 1000 days are a critical window from nutrition point of view. Unfortunately, in several regions of developing countries including India, neither the mother nor the baby is provided with the right quantity and quality of food during this window period. Therefore, it is important for the policy makers to identify malnutrition at an early stage for timely planning and implementation of various intervention strategies at the community, village, and district level. NitiAyog, Government of India (GoI) (previously known as Planning Commission), in an effort to eliminate malnutrition, regularly plans several short and long term nutritional programmes and policies. In this chapter, we will discuss some of these policies launched since independence and their impact in combating malnutrition.

Need of A Nutritional Policy

As per WHO global nutrition policy review report 2010, approximately half of the total deaths worldwide happened due to malnutrition. The root cause of malnutrition in underdeveloped and developing countries is widespread persistence of poverty and hunger. The vicious cycle of poverty as shown in Fig. 14.1 also depicts the same. Poverty hampers the capacity to afford nutritious food and therefore, the person has to compromise with low quality foods and nutrients. This results in under-nutrition and weak immune system. Weak immune system is prone to frequent disease attack; resulting in stunted growth and development. Further, comparatively weak physical, mental and moral strength and impaired productivity, results in low earning capacity for these immune compromised individuals. This cycle goes on for next generation also. Although, poverty is the basis for nutritional problems, yet along with poverty, the other factor responsible for overall nutritional problems in India includes high illiteracy, low production of food grains, low procurement of foods, poor quality of raw food materials, poor utilization of services, false food beliefs, unemployment, poor immunization and health services. Therefore, WHO recommends the implementation of various long and short term nutrition action policies.

Nutritional Problems in India

In India, women (especially pregnant and lactating) and children (6 to 59 months age group) are the most vulnerable group susceptible to malnutrition. Some of the terms related to malnutrition and major nutritional problems are mentioned below:

Malnutrition: Poor nutrition resulting in imbalance in energy and specific micro- or macronutrients either due to relative deficiency or excess of some essential nutrients in diet is known as malnutrition.

Underweight: Body Mass Index (BMI) <18.50 Kg/m^2

Overweight: BMI ≥ 25.00 Kg/m^2

Obese: BMI ≥ 30.00 Kg/m^2

Under-nutrition: Inadequate or insufficient food and nutrients intake for a longer period of time or excessive loss of the same due to some metabolic disorder.

Low birth weight: At birth, if the weight of the child is less than 2.5 kg (5.5 pounds), then that comes under low birth weight.

Anemia: A medical condition in which either the red blood cells (number) or their oxygen carrying capacity is not sufficient enough to meet the physiological needs of the body. These physiological needs may vary with age, sex, pregnancy, smoking etc. Iron deficiency in diet or its poor bioavailability from the diet is the main cause of anemia. However, other factors such as status of vitamin A, B_9, B_{12}, inflammation or any parasitic infection may also play a significant role.

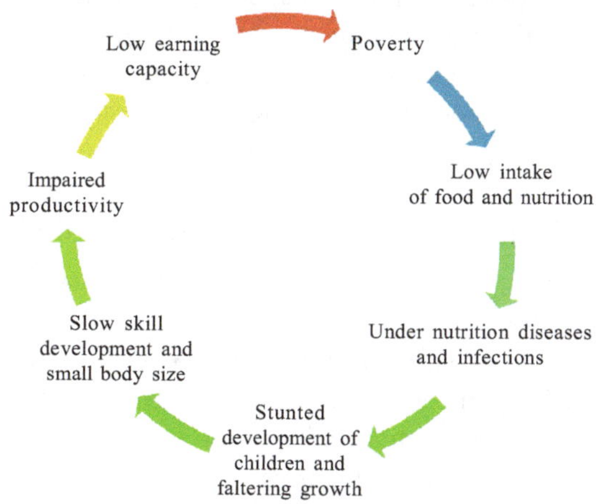

Fig. 14.1: Vicious cycle of poverty

Vitamin A Deficiency: This is a major public health problem in India, characterized by serum retinol level below 0.70μmol/L.

Iodine Deficiency: Iodine deficiency, caused by low dietary iodine especially in hilly areas is one of the prominent causes for brain damage in childhood.

Protein Energy Malnutrition (PEM): It refers to the deficiency of both proteins and energy in diet in association with other medical complications such as diarrhea, respiratory infections, measles and intestinal worms. Poor environmental conditions, big family, poor maternal health and premature termination of breast feeding further exuberates this situation. Marasmus, characterized by extreme wastage of muscles & kwashiorkor, characterized by swelling of the distended abdomen, moon face & dermatitis are the two clinical forms of PEM.

Nutritional Policy Instruments

Combating malnutrition is an issue that could be solved by multi-step approach. Therefore, the problem can be tackled through both direct and indirect interventional approaches. Direct approach is meant especially for the vulnerable group of the population whereas various long term nutrition intervention policies could be a handy tool in long run for improving the nutritional status. Direct approaches include educating the adolescent girls, awareness among mothers regarding nutritional status of babies, warranting improved coverage of expectant women, fortification of essential foods, promotion of inexpensive nutritional foods and control of micro-nutrient shortages among susceptible groups. Whereas indirect approaches include planning for food security (establishment of warehouses, cold stores, godawns), changes in the dietary pattern of a group of population keeping in mind the specific nutritional problems of that area, improvement in the purchasing power through various poverty alleviation programmes, transparency in public distribution system, land reforms, educating people about hygiene, basic health & nutrition and regular surveillance/monitoring of various nutrition programmes/schemes.

Intervention Programmes to Combat Malnutrition

Since independence, Government of India (GoI) planned and implemented various nutrition intervention programmes under different ministries to tackle the problem of malnutrition. Some of these programmes have proved to be great success while some have been reformed from time to time to combat the challenges ahead. The major highlights and salient features of some of these central intervention programmes are discussed below:

(1) Integrated Child Development Services (ICDS)

As per the population census-2011, zero to six years children represents 154 million of Indian population. Therefore, Ministry of Women and Child Development (MWCD) has launched various schemes for prosperity, progress and protection of children. ICDS is also one of these schemes and star programme by GoI, launched on October 2nd 1975. It is one of the world's largest early childhood care and development programme. In addition to 0-6 year old children, pregnant and lactating women are also beneficiaries under this scheme. GoI allocated 16,745 Crore rupees for this programme in the financial year 2018-19.

Objective of the scheme

(a) Improvement in both nutritional and health status of zero to six year old children.

(b) Laying foundation for appropriate psychological, physical and social development of the child.

(c) Diminishing incidences of mortality, morbidity, malnutrition and school dropouts.

(d) Achieving operative co-ordination of policy and execution amongst the various departments to promote child development.

(e) Enhancing ability of the mother to look after the regular healthiness and nutritive needs of the children via appropriate nutrition and health awareness.

ICDS Team

These all services are provided through a platform called 'Anganwadi centres' (AWCs). Currently, 13.46 lakh AWCs under 7076 projects are functional all over the country. ICDS team consists of anganwadi workers, helpers, supervisors, child development project officers and district programme officers. Besides this, the medical officers, auxiliary nurse midwife and accredited social health activist (ASHA) are also a part of the team to accomplish amalgamation of different services.

ICDS also offer services such as supplementary nutrition, pre-school non-formal education and awareness about nutrition & health to children & women. Ministry of Health and Family Welfare (MHFW) provides immunization and vaccination services, regular health checkup and referral services under this scheme.

In this scheme, the supplementary nutrition is provided to the children (0-6 year age group), pregnant women and lactating mothers to fulfill the gap of nutrients between their average daily intake and recommended dietary allowances (RDA). The norms of nutrition and feeds under supplementary nutrition component of this program are revised regularly to combat the malnutrition. Latest revision in these norms took place w.e.f. February 24th, 2009. As per the revised norms, it is mandatory to provide supplementary nutrition for minimum 300 days in a year to the beneficiaries. The mandatory protein and energy requirements as per the revised norms are shown in Table-14.1

Special Nutrition Programme (SNP)

Special Nutrition Programme was launched in the year 1970-71 by Ministry of Social Welfare, GoI with aim to enhance the nutritional status of some specific target groups living in urban slums, tribal areas and backward rural areas. It provides supplementary nutrition for about 300 days in a year with 300 calories and 10-12g of protein/day to preschool children and about 500 calories and 25g of protein/day to expectant and nursing mothers.

Table-14.1: Supplementary nutrition as per revised ICDS norms

Beneficiaries	Calories	Protein (g)
6-72 month children	500	12-15
6-72 month severely malnourished children	800	20-25
Pregnant and lactating women	600	18-20

During the sixth five-year plan, along with providing supplementary nutrition, supply of vitamin A solutions, folic acid tablets and iron tables was also added in to the components of SNP.

Balwadi Nutrition Programme (BNP)

This programme is also operational since 1970-71 and is being implemented by Central Social Welfare Board (CSWB) in association with some non-governmental voluntary organizations such as Indian Council for Child Welfare, Harijan Sevak Sangh, Bhartiya Adimjati Sevak Sangh and Kasturba National Memorial Trust. CSWB provides grants to these voluntary organizations and trusts for implementation and smooth running of the programme. The beneficiaries of this program also include disadvantaged segment of the society similar to SNP. The additional nutrition is delivered to these beneficiaries through BNP. In addition to supplementary nutrition (300 calories and 10g protein/day for 270 days in a year), social and emotional development of beneficiaries is also an objective of BNP.

Wheat Based Supplementary Nutrition Programme (WBSNP)

WBSNP is a scheme launched in 1986 and sponsored by central government. The main aim of this scheme was to enlarge the coverage of existing nutrition programmes by covering additional beneficiaries and popularize wheat intake. Pre-school going children, expecting women, pregnant women and lactating mothers of the areas with high infant mortality rates or areas with higher population of scheduled castes are the main beneficiaries under this scheme. The supplementary nutrition is provided to these beneficiaries under two components as shown in Table-14.2.

Table-14.2: Activities under WNP

Centrally funded component		State funded component
For children	Energy-300Calories & Protein 10g per day	Initially Centre provided wheat to the respective States at a subsidized rate (Rs. 700 per metric ton) However, from1989 no subsidy is provided from Centre to State governments
For expectant and nursing mothers	Energy-500Calories & Protein 20g per day	

Mid-Day Meal Scheme (MDMS)

This scheme was started as National Programme of Nutritional Support to Primary Education (NP-NSPE) on 15th August, 1995. The enhancement in enrollment, withholding and attendance of primary school children along with improved nutritional status were the twin objectives of this scheme. In 2001, the scheme was modified and named as MDMS. According to this scheme, every child in government school or government-aided school would get cooked mid-day meal, meeting nutritional norms (300 calories and 8-12% protein per day) for minimum 200 days in a year. Later on, in 2002, children studying in schemes other than the government school such as education guarantee scheme and alternative and innovative education centers were also included in this scheme. In September, 2004, the scheme was further revised and government's share in cooking and transport cost was increased. As per July, 2006 revision of scheme, the nutritional norms was further increased to 450 calories and 12% protein per day. In October, 2007, the scheme was further extended for the children admitted in class VI to VIII and it was again renamed as National Program of Mid-day Meal in Schools. The nutritional norms for upper primary class children were fixed as 700 calories and 20% protein per day.

National Nutritional Anaemia Prophylaxis Programme (NNAPP)

As discussed earlier in this chapter, anaemia is a serious nutritional problem in India. Therefore, to prevent nutritional anemia in women (expectants, nursing mothers and women acceptors for family planning) and children (1-5 year age group), NNAPP was launched by GoI in 1970 during the fourth plan of planning commission. The women and children with haemoglobin level below 8g/dl & 10g/dl, respectively, were targeted in this scheme. The women and children were provided one tablet of iron + folic acid containing 60mg elemental iron (180mg ferrous sulphate & 0.5mg folic acid) and one tablet of iron + folic acid containing 20mg elemental iron (60mg ferrous sulphate & 0.1mg folic acid) daily for 100 days in a year, respectively. In addition to this, as iron deficiency can be prevented by providing iron through fortification in any suitable food vehicle for 12-18 months, therefore, iron fortified salt was also made available through the scheme.

National Prophylaxis Programme against Nutritional Blindness due to Vitamin A Deficiency (NPPNB)

In the decade of 1950 and 1960s, the country witnessed very high rate of zerophthalmic blindness. Therefore, NPPNB was launched in 1970 with an aim to prevent nutritional blindness caused by severe vitamin A deficiency. Initially, the program was launched in only 11 states. However, in 1976, it was extended to other parts of the nation also on the basis of recommendation of

National Institute of Nutrition (NIN), Hyderabad. In 1994, the programme was shifted under National Child Survival and Safe Motherhood Programme (NCSSM). The age group of eligible children was also revised and restricted to only 9 to 36 months of age. As per the programme guidelines, the children were to be provided five doses of vitamin A before reaching their third birthday. In 2006, based on the recommendation of WHO, UNICEF (The United Nations Children's Fund) and MWCD (The Ministry of Women and Child Development), the age group was again revised to 6 to 59 months. The porgramme proved to be successful one as the clinical vitamin A deficiency was now limited to only few geographical locations. Presently, vitamin A supplementation is carried out through primary health centres, their sub centres and ICDS functionaries.

National Goitre Control Programme (NGCP)

Iodine is a structural component of thyroxine hormone, secreted through thyroid gland. Its daily requirement varies from 100 to 150μg for normal human beings and its deficiency can lead to severe medical complications such as mental retardation, cretinism, still birth, deaf mutism and goiter. Therefore, realizing the severity of consequences and magnitude of the problem, NGCP was launched by Central Government in1962. In 1992, it was renamed as National Iodine Deficiency Disorders Control Programme (NIDDCP). The programme is being executed by MHFW. The main goals of the programme were to check that the household salt used for consumption is iodine fortified and to bring iodine deficiency disorders (IDD) to below 5% in the country. The major objectives of the programme are as mentioned below:

1. Assessments to evaluate the degree of IDD in the districts.
2. Swapping common salt with iodized salt for domestic usage purpose.
3. Reassessments to measure IDDs and the effect of iodized salt after every 5 years in the localities/districts.
4. Laboratory monitoring of iodized salt and urinary iodine elimination.
5. Health Education and Publicity.

NIDDCP has been proved to be effective in reducing total goiter rate significantly all over the country. Food Safety and Standards Authority of India (FSSAI) has also introduced a new regulation (Regulation 2.3.12) in 2011, which enforces prohibition and restriction on sale of common salt for direct human consumption. 35 states/Union territories (UTs) have established IDD control cells and IDD monitoring laboratories in their respective states/UTs. In addition to all this, several information education and communication activities for spreading awareness about IDDs have been regularly carried out through various media platforms such as Doordarshan, All India Radio (AIR), print media, songs and drama.

National Diarrheal Diseases Control Programme (NDDCP)

NDDCP was launched in 1978 with the aim of minimizing the mortality and morbidity due to diarrheal diseases. In 1992, the implementation of this program was handed over to Child Survival and Safe Motherhood Programme (CSSMP). Presently, it comes under National Rural Health Mission (NRHM). Better maternal & child health care practices and strategies for preventing diarrheal epidemics are long term components of this programme. Oral rehydration therapy to prevent diarrhea is one of the major initiatives of this programme as oral intake of glucose increases the intestinal absorption of salt and water, helping in maintaining electrolyte and water balance.

Conclusion

Malnutrition is a multi-dimensional problem, being affected by a number of factors such as poverty, education level, hygiene level, socio-religious restrictions etc. Government had to be proactive in implementation of all schemes. Children, expecting women, pregnant women and lactating mothers are more prone to nutritional deficiencies as they need comparatively more nutrition. Further, as India is a diverse country, the nutritional needs of persons from different communities and different geographical locations vary, therefore, the objectives of the nutritional schemes should be set according to the area specific needs rather than national standards. Awareness programmes about personal hygiene and importance of education should be conducted through visual and print media. There should be a more appropriate, transparent, quick and effective documentation and monitoring system for providing accurate updates of the status of nutritional schemes. Government of India should also increase the share of annual budget for these schemes. Further, more participation of non-government organizations, cooperative societies and trusts in proper implementation of these schemes could prove to be a milestone.

Sample questions

1. Why we need nutritional policies in India.
2. Briefly highlight the nutritional problems in India.
3. Explain the vicious cycle of poverty.
4. Briefly explain direct and indirect nutritional policy instruments.
5. Enlist various nutritional programmes and schemes launched by Govt. of India to combat malnutrition.
6. Highlight the main objectives of Integrated Child Development Services.
7. Briefly describe Special Nutrition Programme and Balwadi Nutrition Programme.
8. When was the programme 'Wheat based Supplementary Nutrition' launched? Highlight the main activities under this programme.
9. What is Mid-Day Meal Scheme?
10. Briefly discuss National Goitre Control Programme and National Diarrheal Disease Control Programme.
11. Discuss about the intiatives of Govt. of India to control anaemia and vitamin A deficiency.

Suggested Readings

Kapil, U. and Sachdev, H.P.S.(2013). Massive dose vitamin A programme in India-Need for a targeted approach. *The Indian journal of medical research*, 138(3), p.411.

Lucas, A., & Sampson, H. A. (Eds.). (2006). Primary prevention by nutrition intervention in infancy and childhood (Vol. 57). Karger Medical and Scientific Publishers.

Rai R.K., Fawzi W.W., Barik, A., and Chowdhury, A. (2018). The burden of iron-deficiency anaemia among women in India: how have iron and folic acid interventions fared? WHO *South-East Asia J Public Health*, 7:18-23

Sahu, S.K., Kumar, S.G., Bhat, B.V., Premarajan, K.C., Sarkar, S., Roy, G. and Joseph, N. (2015). Malnutrition among under-five children in India and strategies for control. *Journal of natural science, biology, and medicine*, *6*(1), p.18.

Seshadri, S. R., & Ramakrishna, J. (2017). Nutritional Adequacy, Diversity and Choice among Primary School Children: Policy and Practice in India. Springer.

WHO (2018). http://www.who.int/news-room/fact-sheets/detail/malnu trition

http://wcd.nic.in/sites/default/files/nnp_0.pdf

http://www.who.int/nutrition/EB128_18_Backgroundpaper1_A_review_of_nutritionpolicies.pdf

http://www.authorstream.com/Presentation/mubarakneo-433665-nutritional-problems-in-india/

http://www.icds-wcd.nic.in/icds.aspx

http://nutrition-health-education.blogspot.com/2014/01/special-nutrition-programme-snp.html

http://nutrition-health-education.blogspot.com/2014/01/wheat-based-supplementary-nutrition.html

http://dghs.gov.in/content/1348_3_NationalIodineDeficiency.aspx

https://www.slideshare.net/nikhilgupta129/diarrhoeal-control-programme

https://www.worldvision.org/hunger-news-stories/top-nine-countries-fighting-child-malnutrition

15

Safety Aspects of Food Additives, Toxic Elements, Radionuclide

Introduction

The dairy sector is one of the most important food sectors, which introduced the physical, chemical, microbiological and sensorial criteria during processing steps and in the final dairy products. Several agents/ingredients of non-dairy origin are commonly used to maintain the quality of milk and milk products during processing, transportation and storage. In general, such ingredients are known as food additives and it can be antioxidants, coloring agents, flavoring compounds, preservatives (naturals and synthetic), artificial sweeteners, vitamins, minerals, stabilizers, emulsifiers and curing agents. From the safety point of view, some toxic agents can come into contact of dairy products during milking, storage or processing; such as fungal toxins, heavy metals, antibiotics or radionuclide's, etc., which need to be identified and processed accordingly. Although addition of food additives in food/dairy product is controlled under country specific food laws or Codex Alimentarius Commission (CAC), few dairy products such as butter and unflavored fermented milk products are highly restricted for food additives.

Food additives

Food additives are substances which are intentionally added to foods to maintain or improve its physico-chemical and microbial quality in terms of freshness, texture, mouthfeel, taste & flavor and sensorial characteristics. Several food additives have been used since a long time such as sugar (in jam, jellies, and syrups), salts (in pickles, fish and meat products) and vinegar. With time, other food additives have also been emerged either natural or synthetic. Common natural additives are color pigments, antioxidants, sweeteners, thickeners and flavor enhancers, etc. derived from plants, animals, insects and microorganisms. Other additives prepared synthetically in labs come under the category of 'synthetic food additives' including artificial sweeteners, flavor enhancers, colours and some antioxidants.

Before addition to foods, additives are needed to ensure the safety and suitability for human health. The addition of any additives is justified only when it performs a particular technological function in food.

Need for food additives

Food additives are intentionally added to foods at a particular concentration due to following reasons:

1. ***To improve or maintain quality, safety & shelf-life of food:*** Several preservatives such as salt, vinegar, sorbic acid, benzoic acid, etc. are added to foods to reduce the growth of the microorganisms and to improve the microbial quality of food during transportation and storage. A few antioxidants such as butylated hydroxyl anisole (BHA), butylated hydroxyl toluene (BHT), ascorbic acid, etc. are added to fat rich foods to inhibit the oxidation of fats and production of rancid/off-flavors compounds.
2. ***To improve or maintain the nutritional quality of food:*** A wide range of food products are supplemented with several vitamins and minerals (additives)such as vitamin C, fat soluble vitamins (A, D, E) and Vitamin $B_{12,}$ Na, K, Ca, Fe, Zn, etc. to compensate the amount in diet which otherwise lost during processing and storage of food.
3. ***To improve the appearance, taste and texture:*** Nowadays, color addition has become very common in food and dairy products as it improves the appearance of the products. Several emulsifiers and stabilizers are added to improve the viscosity, emulsion stability, mouth feel and texture of the product. A few ingredients are added as fat replacers to develop fat-free or reduced fat products. Artificial sweeteners and flavours are added to improve the taste and flavor of the product.

Direct *vs.* Indirect food additives

Food additives directly added to foods for a specific function/s are known as direct food additives. For example, thickeners/modified starches added to salad dressings, bakery products, mayonnaise, etc. Mostly direct additives are labeled on packaging material. Whereas, indirect food additives are not intentionally added into food and become the part of it during processing, storage and/or coming into contact with packaging material. For example, Fe or other metal ions from machinery, plastic polymers/colours from the packaging materials.

More than 3000 food additives have been identified till date, which have been studied and tested for their safety and suitability for humans by Food and Drugs

Administration (FDA). Food additives have been categorized into different groups, which are discussed below:

Antioxidants

Antioxidants are chemical molecules which inhibit the oxidation of other molecules in living or non-living system. Oxidation is a process in which electrons or hydrogen ions transfer from a molecule to the oxidizing agent, resulting oxidized molecules and free radicals generation. In food system, these free radicals initiate chain reactions and favor the production of more and more free radicals and oxidized molecules. Oxidation is a common problem in fat-rich foods, where fat is oxidized by atmospheric oxygen and generates free radicals, which ultimately causes the production of hydroperoxides, also known as primary oxidation products. These products are toxic for humans. Antioxidants scavenge these free radicals in foods and inhibit the formation of further free radicals and oxidized products. In other words, antioxidants are strong reducing agents, which prevent oxidation by being oxidized themselves.

On the basis of solubility, antioxidants can be divided into hydrophilic (water soluble) and hydrophobic (fat soluble) antioxidants. However, on the basis of nature, antioxidants can be divided into 'Natural' and 'Synthetic' antioxidants. For example, vitamins C, citric acid, phytic acid, phenolic acids, tocopherols, flavonoids, lycopenes, etc. are natural antioxidants. While, BHA, BHT, tertiary butyl hydroquinone (TBHQ), propyl gallate (PG) are synthetic antioxidants. Generally, synthetic antioxidants are added in fat rich foods to prevent the fat

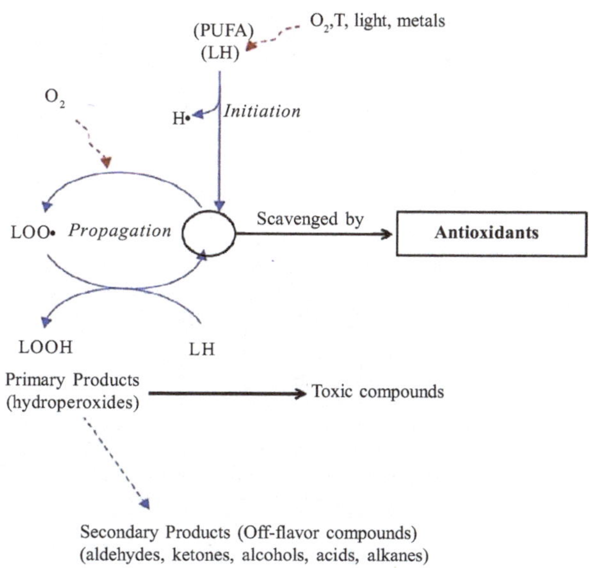

Fig. 15.1: Mechanism of oxidation of fat in presence of oxygen, heat or metal ions

Table 15.1: List of different antioxidants, their ADI, functions and food applications

Antioxidant	ADI(mg/kg body weight)	Nature/Function	Applications
Ascorbic acid	Not limited	Natural antioxidant, enzyme inhibitor, reduce discoloration	Beers, cut fruits and vegetables, dried potatoes, jams, jellies, etc.
Citric acid	Not limited	Natural antioxidant, enzyme inhibitor, reduce discoloration	Biscuits, jams, jellies, canned fruits, alcoholic beverages, fruit juices, cheese, etc.
Butylated hydroanisole (BHA)	0-0.5	Synthetic antioxidant, prevent fat oxidation	Fats, vegetable oils, fat rich snacks and products
Butylated hydroxytoluene (BHT)	0-0.3	Synthetic antioxidant, prevent fat oxidation	Fats, vegetable oils, fat rich snacks and products
Tertiary-butylated hydroquinone (TBHQ)	0-0.2	Synthetic antioxidant, prevent fat oxidation	Fats, vegetable oils, fat rich snacks and products
Sulphites (As sulphur dioxide)	0-0.7	Synthetic antioxidant, to prevent spoilage and oxidation	Wine, dried fruits, canned vegetables, frozen foods

oxidation. Antioxidants can be naturally present in foods or may be intentionally added. Fruits particularly, citrus fruits are rich in antioxidants. The mechanism of oxidation of fats is shown below (Fig. 15.1):

Acceptable daily intake (ADI)

ADI of common antioxidants along with their major applications is shown below (Table 15.1):

All natural antioxidants have the GRAS (Generally recognized as safe) status, for which maximum permissible levels are not specified by regulatory agencies. However, synthetic antioxidants have some restrictions and are not allowed in all the products. According to the food additive regulations given by FDA, BHA and BHT can be added individually orin combination at a maximum level of 0.02% (200 ppm), based on the lipid content of food products. For example, BHA and BHT are not allowed in fish products. Although BHA and BHT are effective at low concentrations, they become pro-oxidant (induce oxidation) at high levels in foods. Chemical structures of commonly used antioxidants are shown in Fig. 15.2.

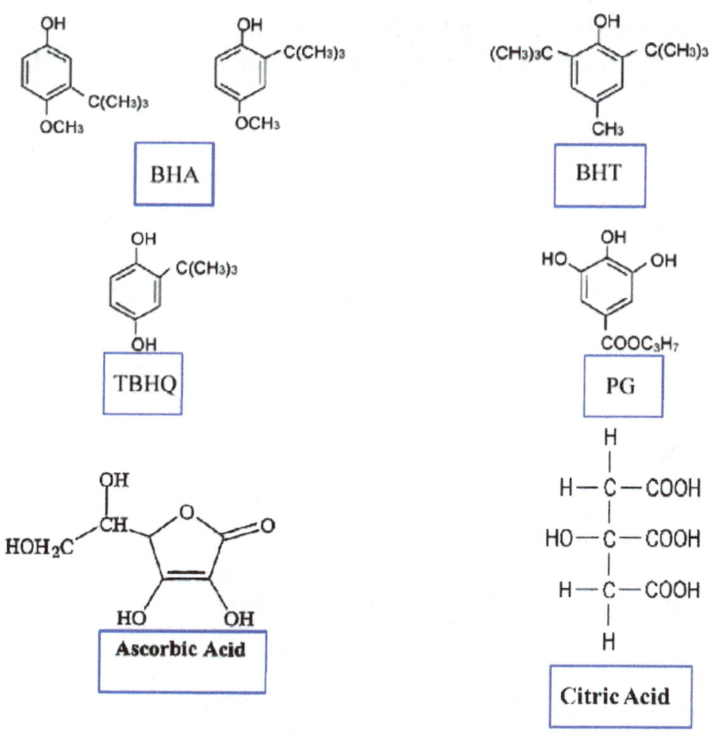

Fig. 15.2: Chemical structures of commonly used antioxidants

Chelating agents

Several metal ions such as iron, copper, zinc, chromium, etc. are naturally present in several foods in complex forms, e.g., Mg in chlorophyll, Fe in porphyry in rings, copper complexed with amino acids, etc. These metal ions, in human body, are essential for enzymatic activity. However, these metal ions can be released in foods from its complex form during processing conditions, such as heating, sterilization, acidification, fermentation, etc. Free metal ions are free to participate in several chemical reactions that cause loss of eye appeal (appearance), mouth feel, texture, taste & flavor, nutritional value and overall consumer acceptability of foods. Sometimes, free metal ions react with high molecular weight compounds such as proteins, tannins, phytic acids, etc. and form insoluble complexes. Therefore, the compounds which are added in foods in minute amount during processing of foods for chelating (binding) free metal ions are known as chelating agents or sequestrants.

Chelating agents can also bind with the metal ions already complexed with enzymes. A few enzymes are responsible for spoilage of foods during processing and storage and require metal ions for their activity. Chelating agents bind with those metal ions and reduce the enzymatic activity and thus, improves the stability of foods. In this way, it promotes the colour retention in dried foods and flavor retention in canned fruits & vegetables, carbonated beverages, salad dressings, mayonnaise, margarine and sauces.

It should be noted that chelating agents are not antioxidants, but they show synergistic effects and help the antioxidants in their action. Free metal ions are also responsible for the oxidation of fats and oils, thus added chelating agents can bind metal ions and reduce the chances of oxidation. In other words, chelating agents not only bind with the free metal ions, but also improves the efficiency of antioxidants.

Compounds containing two or more functional groups such as hydroxyl, sulphydryl, carboxyl and phosphates can act as chelating agents. The examples are given below:

a. Polycarboxylic acids (citric, malic, tartaric, oxalic and succinic acids)

b. Sorbic acid

c. Polyphosphates (ATP and pyrophosphates)

d. Macromolecules (Proteins, porphyrins)

e. EDTA (Ethylenediaminetetraacetic acid) (Fig. 15.3)

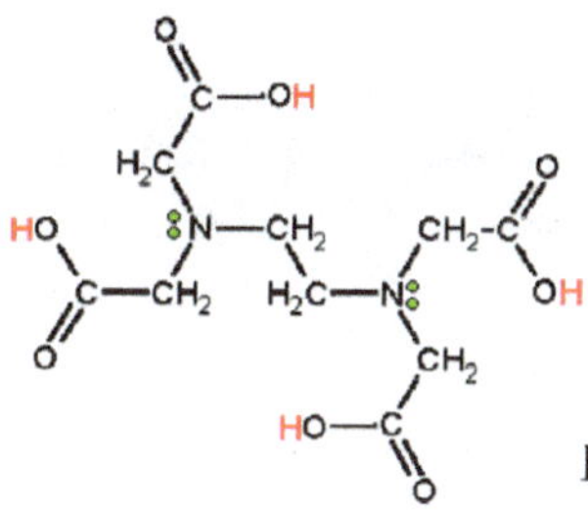

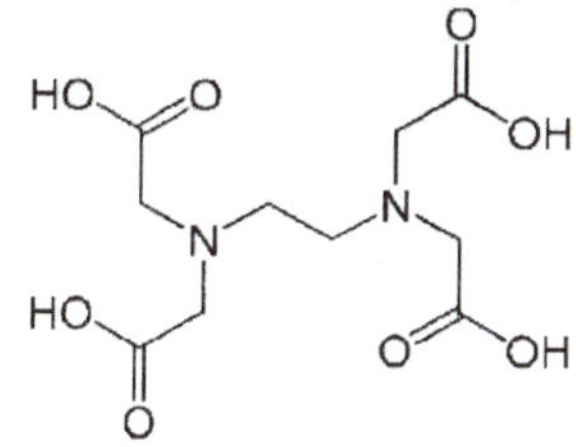

EDTA

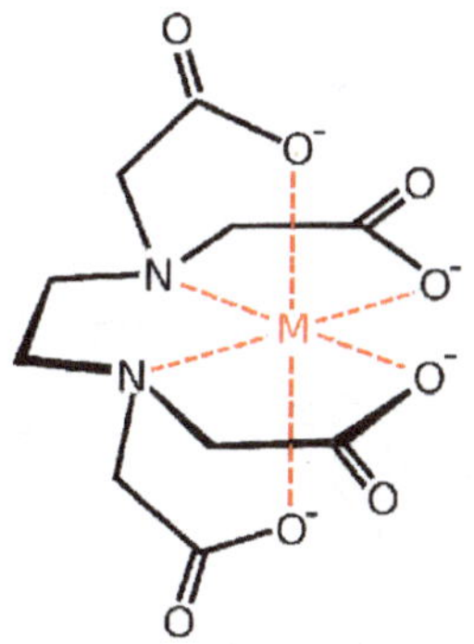

EDTA complex with metal ion

Fig. 15.3: Chemical structure of EDTA

Though amino acids, polyphosphates, organic acids can chelate the metal ions, but they form weaker complexes and thus, are less effective. However, EDTA is the most common and versatile synthetic chelating agent added as food additive as it forms strong complex with metals.

When dissolved in water in the form of disodium EDTA (Na_2H_2EDTA), EDTA molecules can form highly stable complex ions with two- and three-valent metal ions. EDTA is generally added in foods as the disodium salt of its calcium complex ($Na_2CaEDTA$) due to its more solubility in water.

How EDTA works?

Normally EDTA works by chelating metal ions in foods. In emulsions such as salad dressings, mayonnaise and margarine, EDTA works as shown below (Fig. 15.4):

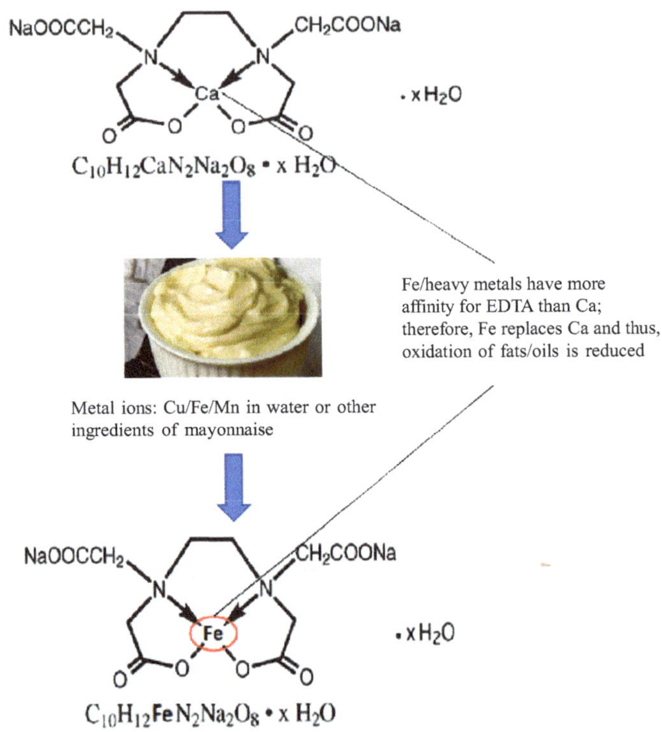

Fig. 15.4: Working mechanism of EDTA

Common chelating agents along with their ADI, maximum permissible level in foods, applications and functions are given below (Table 15.2):

Table 15.2: List of commonly used chelating agents along with their ADI and food applications

Chelating agent	ADI (mg/kg body weight)	Applications	Maximum permissible level in foods
EDTA ($Na_2CaEDTA$)	0-2.5	Mayonnaise, margarine and fatty emulsions, beverages, canned fruits and vegetables, dried fruits/nuts, etc.	0.05%
Citric acid	Not limited	Bakery products, fruits juices, beverages, canned fruits and vegetables, etc.	0.1-0.3%
Polyphosphates	70	Cut fruits and vegetables, soft drinks, meat and fish products	0.5%

Functions of various chelating agents

1. Used as an iron fortificant, particularly EDTA
2. Inhibits fat oxidation in fat rich products and dairy/non-dairy emulsions
3. Inhibits polyphenol oxidase by reducing pH and thus reduce browning of cut fruits and vegetables
4. Reduce discoloration in acidic beverages by chelating metal ions, reduce oxidative and flavor changes in soft drinks
5. Reduce cloudiness in fermented malt beverages by complexing with Cu, which otherwise catalyze the oxidation of phenolic compounds and interact with proteins and cause turbidity
6. Polyphosphates improve water holding capacity of muscles protein in meat and fish products; reduce discoloration of sea food by chelating copper ions and inhibits fat oxidation

Colorants

As per Food and Drug Administration (FDA), colorants or color additive is any dye, pigment or substance that can impart colour, alone or through reaction with other substances, when added or applied to a food, drug and cosmetic or to the human body. Colorants are also known as food dyes, coloring agents or food colorants.

Why colorants are added in foods?

Colorants may be added in food products for the following purposes:

1. To replace the original color lost during the processing steps
2. To improve the appearance/eye appeal of the product
3. To mask variations of original colour in different batches
4. To enhance naturally occurring colors
5. To protect photo-sensitive flavors and vitamins in foods
6. For decorative or artistic purpose, e.g. in cakes, pastries
7. Sometimes to mask defects in foods
8. To color otherwise uncolored product

Classification of colorants

On the basis of origin, color additives can be divided into two categories: 1) Natural, 2) Artificial (Synthetic). Natural colour additives can be obtained from plants, animals, microorganisms and/or minerals. Artificial colorants are dyes obtained through chemical reactions and generally inexpensive when compared to natural colors. As per US FDA, colorants can be divided into **'Exempted (Non-certified)'** and **'Certified'** categories.

Exempted colors include pigments extracted from natural sources such as vegetables, minerals, microorganism or animals. For example, annatto extract (yellow), dehydrated beets (bluish-red to brown), caramel (yellow to tan), beta-carotene (yellow to orange) and grape skin extract (red, green). Although exempted color additives are not subject to batch certification requirements, they are still color additives and FDA must approve them before they can be used in foods.

Certified color additives are synthetic colorings that are inexpensive, used widely; impart uniform and intense colour and water soluble. These additives are classified as certified because they are required to undergo certification every time a new batch is manufactured. Detailed classification of colorants with examples has been shown below (Fig. 15.5):

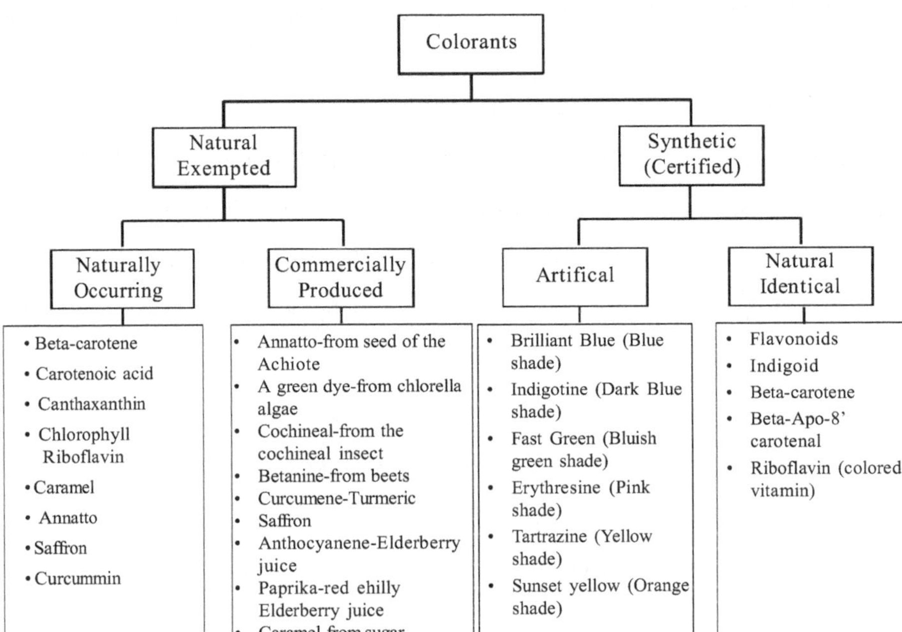

Fig. 15.5: Detailed classification of colorants with examples

Naturally occurring colorants: Naturally occurring colorants are color additives extracted or isolated from natural sources. A list of natural colorants in given in Table 15.3.

Commercially produced colorants: Commercially produced colorants are natural molecules/pigments but produced commercially. There is minor difference between naturally occurring and commercially produced natural colorants. Natural colorants are generally fat soluble and least water soluble.

Artificial colorants: Artificial colorants are synthetic compounds/dyes, not present in nature and obtained through chemical reactions. They are generally inexpensive, water soluble and can be used in foods without any further processing. Before use in foods, they should be approved by FDA.

Nature identical: Natural colors, their extraction and purification are costly. To overcome this, identical molecules have been prepared in labs by chemists, known as nature identical colorants. In other words, nature identical colors are exactly the same molecules found in natural sources but they are made synthetically.

Regulations for Colorants

All the color additives are regulated by UD FDA. FDA ensures the safety and suitability of each colorant for human health. Usage of synthetic colors has been restricted to a maximum limit of 100-200ppm. ADI values of synthetic dyes/colorants approved by FDA are shown below (Table 15.4):

Table 15.3: List of natural colorants, sources, colour and food applications

Chemical classification	Colorants/additive	Sources	Color	Food applications
Carotenoids	Beta-carotene	Carrots (natural) or synthesized (nature identical)	Yellow to orange to red	Bakery products, fruit juices, beverages, dairy products, canned fruits and vegetables, etc.
	Canthaxanthin	Synthesized (nature identical)		
	Paprika (oleoresin)	Paprika (*Capsicum annum* L.)		
	Annatto	*Bixa orellana* L.		
	Astaxanthin	Synthesized, bacterium (*Paracoccus carotinifaciens*), yeast (*Phaffia rhodozyma*), algae (*Haematococcus pluvialis*)		
	Lycopene	Tomato (*Solanum lycopersicum*)		
Caramel	Caramel	Heated sugars	Brown	General
Anthocyanin	Grape skin extract	Grapes	Red to purple to blue	Beverages, general purpose
	Fruit juices	Berries		
	Vegetable juices	Carrot, cabbage, others		
Betalain	Beet root powder	Beetroot	Pink to red	General purpose
Minerals	Iron oxide	Synthesized	Yellow-orange-red-brown-black	Confections, pet food
	Titanium dioxide	Synthesized	White	General
	Ferrous gluconate	Synthesized	Yellowish-gray	Ripen olives
	Ferrous lactate	Synthesized	Greenish white	Ripen olives

Table 15.4: List of artificial colorants, their color and ADI value

Color-Artificial colorant	ADI (mg/kg body weight)
Red colour-Ponceau 4R	4.0
Red colour-Carmoisine	4.0
Red colour-Erythrosine	0.1
Yellow colour-Tartrazine	7.5
Yellow colour-Sunset yellow	2.5
Blue colour-Indigo carmine	5.0
Blue colour-Brilliant blue	12.5
Green-Fast green	25.0

Curing agents

Curing: Curing is a type of preservation method in which salt, sodium nitrate/ sodium nitrite, sugar etc. are added in foods to extend the shelf life, to enhance flavor and to improve colour during storage. Most common curing agents are sodium chloride, sodium nitrate, phosphates, etc. which are added in the preservation of meat and fish products and vegetables.

Functions of the curing agents

Sodium chloride

1. Inhibit the growth of microorganisms (antimicrobial) by reducing the water activity
2. Enhances the flavor

Sugar

1. Counteracts the harshness of saltby preventing the excessive removal of moisture and by moderating the flavor
2. Sugar provides energy for bacteria that change nitrate to nitrite

Sodium nitrate/nitrite

1. Color fixation (stable pink color)
2. Contribute to the characteristic flavor
3. Inhibit the growth of food poisoning and spoilage bacteria, especially of *Clostridium botulinum*
4. Retards the fat oxidation

Sodium nitrate and sodium nitrite have been used since a long time for curing meat and fish products. The mechanism of working sodium nitrate or sodium nitrite is shown below. After heating or cooking of meat added with curing agent, nitrosylmyoglobin is converted into denatured **nitrosyl-hemochrome** which is of **pink colour (desirable)** and stable during storage. However, during prolonged storage under atmospheric oxygen, further oxidation takes place and ferrous state is converted into ferric state leading to the formation of **brown colored** pigment in meat/food product, which is **non-desirable** and reduces the consumer acceptability (Fig. 15.6).Therefore, cured meats are packaged so that they are minimally exposed to the deleterious effects of oxygen and light.

It has been reported that when nitrite added cured meat is ingested; N-nitrosamines are formed under highly acidic gastric conditions, which are potent carcinogens. Therefore, government has banned/restricted the use of sodium nitrates/nitrites for the curing of meat and fish products.

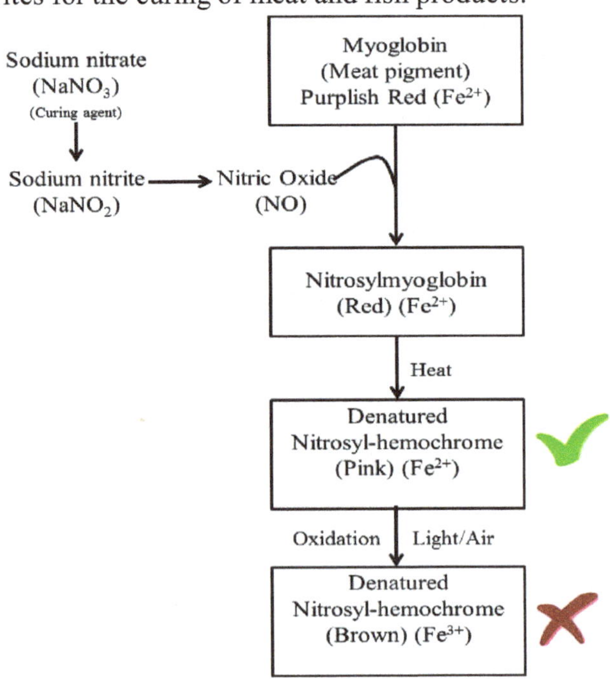

Fig. 15.6: Working of curing agent (nitrites/nitrates) in meat and meat products

Role of sodium ascorbates and isoascorbates in curing

L-ascorbic acid or its sodium salt and isoascorbates are not naturally present in meat products in appreciable amount and are added into it as **curing adjuncts.** They increase the effect of curing agents and facilitate the formation of nitric oxide (NO). Thus, ascorbates and its salts, accelerate the formation of stable pink color by maintaining the reduced conditions on the surface of the product during curing. Even more important role of ascorbates in cured meat is to block the formation of carcinogenic N-nitrosamines.

Role of phosphates in curing

Phosphates are added in meat products during curing to improve the juiciness of muscles' proteins. Phosphates improve the water binding capacity of muscles' proteins by raising the pH. At higher pH, unfolding of muscles protein is increased, thereby making more sites available for binding with water. Due to their antioxidant activity, phosphates also reduce the chances of fat oxidation and thus, help in improving color and flavor.

Flavorings and flavor enhancers

Flavours are low molecular weight (<300 g/mol) volatile organic compounds. Most of the flavoring compounds have well established structures with one or more functional group such as lactones, ketones, aldehydes, esters, alcohols, aliphatic acyclic alcohols, carboxylic acids, etc. For any food, flavor is the most significant attribute affecting the consumer acceptability. Sometimes, during processing, transportation and storage, flavoring compounds are either lost or its intensity is reduced. Therefore, food industries use concentrated forms of flavoring compounds to compensate the loss of original flavor; or for developing new flavored products.

'Flavorings' are the concentrated preparations which are added to foods in order to impart taste and/or flavor. Flavorings are not intended to be consumes as such and must be added in foods depending upon the type of flavoring. Flavorings can be divided into three categories:

1. Natural flavorings/flavors
2. Artificial flavorings/flavors
3. Nature-identical flavorings

Natural flavorings: Flavoring substances that are obtained from natural sources such as plants, animals or microorganisms by physical, microbiological or enzymatic methods are known as natural flavoring substances. Examples: spices, fruit juices, herbs, vegetables, etc.

Artificial flavorings: Flavoring substances that are not obtained from natural sources or identified in a natural product are known as artificial flavorings. These flavorings are synthetically produced in labs using fractional distillation or by some other techniques. Example: Vanillin, the main component of vanilla beans which can be produced as a natural or a nature-identical flavouring substance.

Nature-identical flavorings: Nature-identical substances are the flavoring substances that are obtained by synthesis or are isolated through chemical processes, which are chemically identical to natural flavoring substances. Example: Ethyl vanillin, which is artificial, and smells & tastes like vanillin yet is roughly three times more taste-intensive when added to ice cream, confectionery and baked goods.

A table is shown below showing the chemicals commonly used for flavorings (Table 15.5):

Table 15.5: List of commonly used chemicals and their flavors

Chemical	Flavor
Allylpyrazine	Roasted nut
Methoxypyrazines	Earthy vegetables
2-Isobutyl-3 Methoxypyrazine	Green pepper
Acetyl-L-Pyrazines	Popcorn
2-Acetoxy Pyrazine	Toasted flavors
Aldehydes	Fruity, green
Alcohols	Bitter, medicinal
Esters	Fruity
Ketones	Butter, caramel
Pyrazines	Brown, burnt, caramel
Phenolics	Medicinal, smokey
Terpenoids	Citrus, piney

Source: http://www.foodadditivesworld.com/flavorings.html

Flavor enhancers

The flavor of any food is composed of three attributes: taste, odor (aroma) and trigeminal perception in oral and nasal cavity. There are some components which are capable of modifying, supplementing, increasing or decreasing the original flavor of foods, although they have little or no flavor of their own at typical usage levels. These substances are commonly known as "flavor enhancers" or "flavor potentiators". According to the International Organization of the Flavor Industry, flavor enhancer is a substance with little or no odor at the level used, the primary purpose of which is to increase the flavor effect of certain food components well beyond any flavor contributed directly by the substance itself.

Common food flavor enhancers are:

1. Monosodium glutamate (MSG): MSG is a natural flavor enhancer originally prepared from gluten protein. MSG is a sodium salt of L (+)glutamic acid, most commonly used as food enhancer and permitted world-wide (Fig. 15.7). Other salts like Ca, Mg or K of glutamic acid are occasionally used and not permitted everywhere. The taste of MSG has been described as sweet-salty. Some have described the taste impression of MSG as savory, beefy, brothy, meaty and mouthwatering. The detection threshold of MSG in aqueous solution has been reported to be 100 to 300ppm. The term **'Umami'** was proposed for the flavor-improving effect of MSG, which means deliciousness, palatability, savoriness or succulence in Japanese. Only the completely dissociated form of glutamic acid exerts the umami effect. The dissociation of MSG is pH dependent and imparts umami taste between pH 6-8.

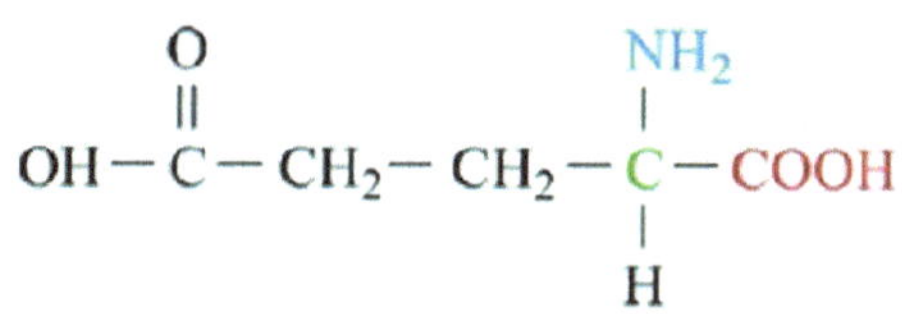

L(+) Glutamic acid

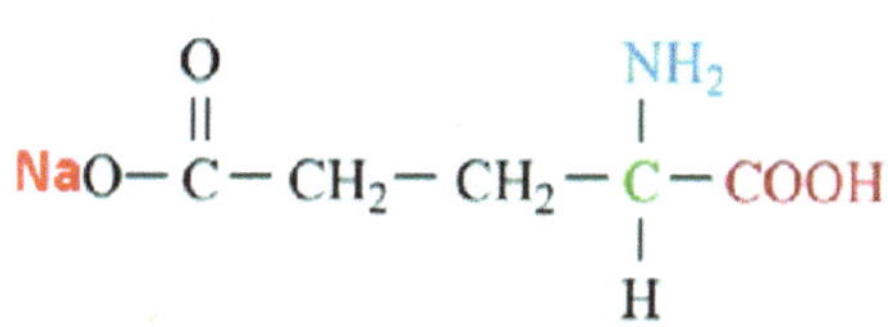

Monosodium glutamate (MSG)

Fig. 15.7: Chemical structures of glutamic acid and monosodium glutamate (MSG)

MSG is commonly used in snacks, soups, ready-to-eat food products, meat & fish products, vegetables, etc. It should be noted that MSG has no flavor enhancing effect in several food products such as dairy foods, fruit juices, carbonated soft drinks, confectionary and desserts, probably due to least dissociation of glutamic acid (~3%) at acidic/lower pH. Glutamic acid is also naturally present in several foods in protein-bound form, but this bound state again does not show any umami taste.

MSG and Chinese Restaurant Syndrome

In late 1960s, it was reported that MSG can cause hypersensitive reactions in some people, for which the symptoms were chest pain, burning sensation in the back of the neck, sweating, nausea, headache, weakness and thirst, known as Chines restaurant syndrome. However, later on after extensive research, it was found that MSG is not the only agent for hypersensitive reactions and allergenic proteins; preservatives, histamine, coloring agents, etc. could also cause the symptoms. Therefore, US FDA placed MSG on the list of food additives having GRAS status and permitted its use world-wide with proper labeling conditions.

2. Purine 5'-Ribonucleotides: 5'-ribonucleotides containing purine bases only can act as flavor modifier/enhancers. The most important flavor enhancers under this category are inosine 5'-monophosphate (IMP) and guanosine 5'-monophosphate (GMP). In order to show the flavor-enhancing effect, the 5'-ribonucleotide must have a hydroxyl group or an amino group in the 6-position of the purine base, as shown in figure below (Fig. 15.8). Another requirement for flavor enhancing effect is the phosphate group at 5' position of ribose sugar. If it is present on 2' or 3' position, flavor enhancing effect is not observed.

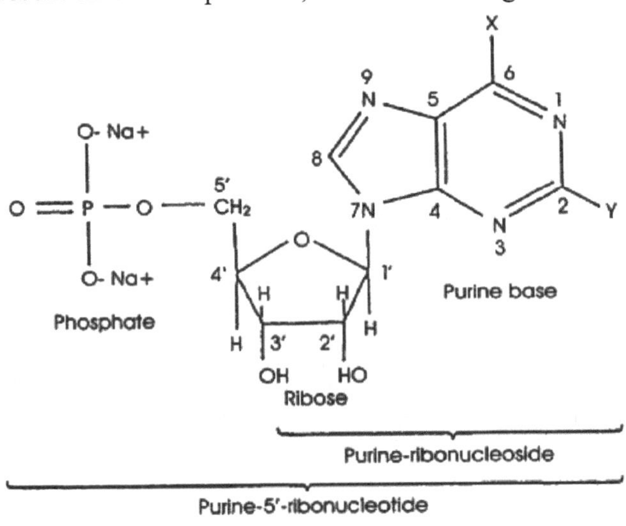

X = OH, Y = H: Inosine monophosphate (IMP) = disodium-5'-inosinate
X = OH, Y = NH_2: Guanosine monophosphate (GMP) = disodium-5'-guanylate
X = NH_2, Y = H: Adenosine monophosphate (AMP) = disodium-5'-adenylate

Fig. 15.8: Examples of disodium salts of 5'-ribonucleotide.

(*Source*: Matheis G. (1997), Flavor Modifiers, In: Food Flavourings. Ed. P. R. Ashurst, Aspen Publishers, pp. 367-421).

IMP and GMP are heat stable (<120°C) and shows optimum flavor enhancing effects at pH 5-7. Taste thresholds in aqueous solution have been reported to be 25-250ppm and 12-200ppm for IMP and GMP, respectively. The taste of IMP and GMP has been described as beefy and oak-mushroom, respectively. IMP and GMP are commonly used in sauces, ketch-up, ready-to-eat foods, meat and fish products, vegetables, etc. at concentration of 0.001 to 0.150% in final food products.

3. Maltol and Ethyl maltol: Maltol is a flavor enhancer commonly used in sweet foods at the levels of 5-75 ppm. It is commonly present in cocoa, coffee, condensed milk, baked products, chocolate, etc. Maltol is formed during heating of carbohydrates during processing of foods.

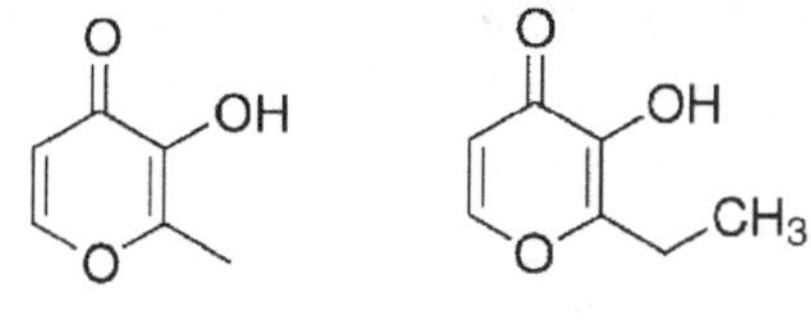

Maltol   Ethyl maltol

Ethyl maltol is a flavor enhancer, sweetener and flavoring agent, not commonly found in foods. It has 4-6 times potency than maltol and generally used in baked goods, chewing gum, jam, chocolate, desserts, ice-cream and other confectionary products. Joint FAO/WHO Expert Committee on Food Additives (JECFA) has recommended up to 2 mg/kg body weight/day as an acceptable limit of maltol and ethyl maltol both for human consumption.

Other flavor enhancers are gymnemic acid, miraculin, curculin, ziziphin, thaumatin, lecithin, etc. which have minor industrial applications.

Preservatives

Any substance which increases the keeping quality and shelf life of food by retarding, inhibiting or preventing the growth of microorganism or any other deteriorative chemical reaction responsible for spoilage of food, is known as preservative. Preservatives can be chemical or biological/microbial origin; natural or synthetic. Here, in this section, we will discuss only chemical preservatives such as salt, sugar, vinegar, acids, etc. Chemical preservatives are added in foods in small quantities (up to 0.2%) and generally do not change the visual or physico-chemical or sensory properties of foods.

The detail of commonly used chemical preservatives with their mode of actions is discussed below:

1. Sodium chloride (NaCl): Salt is one of the oldest preservative agents, still used in a variety of food products. Now-a-days, due to advanced storage technologies, salt is generally used to enhance the taste and flavor of the food products. Salt bind with free water available in food and thus reduce the water activity required for the growth of microorganisms. Microorganisms require water for their enzymatic activity, metabolism and thus survival. Salt draws out the water from the cells and tissues and make it unavailable for the metabolism. Salt is also used for dehydration purpose of several foods such as pickles, snacks, fish, etc. NaCl can be used as brine solution or in power form for preservative action. ADI- not limited.

2. Sugar: Sugar, particularly sucrose, is also one of the most common preservatives used in those products in which it is normally used. Sugar shows its preservative action on the same principle as salt does. The high osmotic pressure of sugar creates unfavorable conditions for the growth of microorganisms. It binds with the free water of food and reduces the water activity of the product. Jam, jellies, syrups, sweetened condensed milk, preserves, confectionary products, marmalades, etc. are a variety of food products where sugar acts as a preservative agent.

3. Lactic acid: Lactic acid is a natural preservative, commonly produced during fermentation of foods from lactose. Lactic acid reduces the pH of the system and does not allow the growth of spoilage bacteria such as putrefactive anaerobes and butyric-acid-producing bacteria. Only health beneficial bacteria can grow at lower pH such as lactic acid bacteria. Although yeast and molds can also grow at lower pH, but by the addition of other chemical preservatives such as sorbic acid, benzoic acid, etc. their growth could be controlled.

4. Acetic acid: Acetic acid can be added directly in the form of vinegar or sodium/potassium or calcium salts in foods for the preservative action against many species of bacteria, yeast and to some extent molds. Acetic acid is more potent preservative than lactic acid and potency further increases on decreasing the pH. Acetic acid in the form of vinegar (4% acetic acid) is commonly used in pickles, vegetables, ketch-up, tomato sauce, etc. Salts of acetic acid are used in bakery products such as bread for controlling roppiness and growth of molds.

5. Citric acid, malic acid, ascorbic acid: All these acids are food-grade preservatives/food additives and are commonly used in a variety of food products such as fruit juices, beverages, preserves, jam, jellies, pickles, etc. Their mode of action is similar to lactic acid.

6. Benzoic acid: Benzoic acid is a lipophilic acid and naturally present is some fruits (like blueberries and plums) and spices (cinnamon and cloves). However, for the industrial use, it is prepared synthetically. Due to its poor solubility in water, benzoic acid is generally used as benzoates (sodium salts), which are highly water soluble (550–630g/L at 20°C). They are effective particularly against yeast and to some extents bacteria.

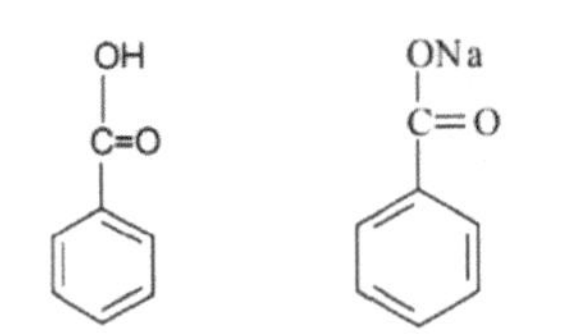

Benzoic acid Sodium benzoate

Benzoates are least active against molds. Benzoates are commonly used in acidic foods such as soft drinks, fruit juices, non-alcoholic beverages, chemically leavened baked goods, condiment, syrups, pickles, sauerkraut, jam, marmalades, etc., preferably in the pH range below 4.5. Due to their inhibitory effect on yeast, they cannot be used in yeast-leavened products. Although undissociated form of benzoic acid is more effective against yeast and bacteria, sodium benzoate is the preferable form having 200 times more water solubility than benzoic acid. It is generally used at the concentration of 0.1% in foods. JECFA recommended the ADI levels of benzoic acid and sodium benzoate from 0 to 5mg/kg body weight/day.

Benzoic acid or its salts show their antimicrobial action by interfering cell membrane components resulting in its destabilization, metabolic enzymes activity particularly of electron transport chain (ETC) and enzymes of oxidative phosphorylation.

*7. **Sorbic acid:*** Sorbic acid [(2E,4E)-hexa-2,4-dienoic acid] is tasteless, odorless and a free flowing powder. It is slightly soluble in water and soluble in ethanol or organic solvents.

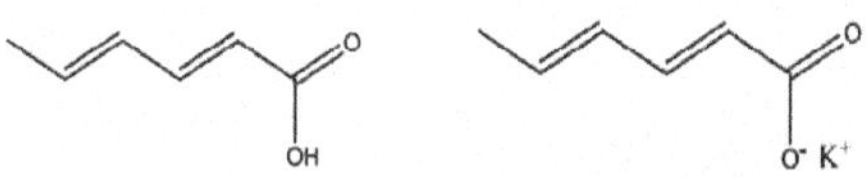

Sorbic acid Potassium sorbate

Unlike sorbic acid, its potassium salt is highly soluble in water as well as ethanol and this is the most commonly used form in food industry. It works optimum in acidic pH ranging from 5-6. As it is non-toxic and completely metabolized; WHO-JECFA has prescribed its ADI 0-25 mg/kg body weight/day, which is the highest among all the chemical preservatives. Major applications of sorbets lies in bakery products, cheese, fruits and vegetables, fruit beverages, pickles, sauerkraut, jellies, jams, preserves, high moisture dehydrated fruits, etc. Like benzoic acid, only undissociated form of sorbic acid has the antimicrobial action. It is approved word-wide and have GRAS status. Other salts such as sodium and calcium-sorbates are also permitted world-wide, but generally sodium salt is not used in food industry due to its instability.

Sorbic acid/sorbates are more effective against yeast and molds as compared to bacteria. Several mechanisms for their antimicrobial action are given below:

- By inhibiting several enzymes of carbohydrate metabolism such as enolase, lactate dehydrogenase, enzymes of electron transport chain, etc.
- By perforation (holes generation) of cell membranes, thus destabilizing the membrane proteins and membrane integrity
- By impairing catalase and peroxidase system of microorganisms
- By inhibiting the absorption of several amino acids
- By inhibiting the multiplication of microorganisms and extending the lag phase

*8. **Propionic acid:*** Propionic acid is a colorless, oily liquid with a pungent, rancid odor. It is naturally present in several dairy products where fermentation occurs and is a by-product of human metabolism. It is prepared synthetically as a food preservative particularly against mold growth. Propionic acid shows its preservative action probably due to inability of the microorganisms to metabolize 3-carbon unit. Its sodium and potassium salts are commonly used for the preservation of cheese and bakery products, where mold growth is a common problem. Optimum activity of propionic acid has been observed up to 5 pH. Propionic acid is listed as GRAS as a chemical preservative and flavoring agents by US FDA. JECFA has not specified its ADI considering it as a normal metabolite in human metabolism and normal ingredient in various food products.

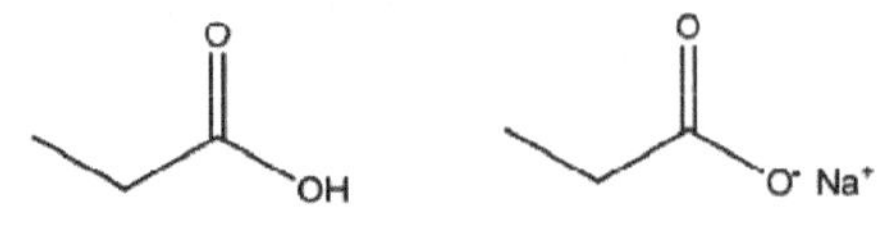

Propionic acid   Sodium propionate

9. Sulphur dioxide: When used as a food additive for preservation purpose, sulphur dioxide is considered a collective term for sulphur dioxide (SO_2), hydrogen sulphite (bisulphite) ions ($HSO3^-$), disulphite (metabisulphite) ion ($S_2O_5^{2-}$) and sulphite ion (SO_3^{2-}). SO_2 is a colorless gas with a characteristic pungent odor. When applied on the surface of food products, SO_2 may be converted in any form/species mentioned above depending upon the pH of the food, its concentration and moisture content. All the forms show equivalent preservative action specifically against yeast and molds except free SO_2, which is the potent inhibitor of microbial growth among its other forms due to its ability to penetrate cell membrane easily. Therefore, this additive is more effective as antimicrobial agent at low pH, where much of it is present in the form of free $SO_{2.}$

Functions of SO_2 and other forms along with their mode of action

i. Sulphur dioxide retards enzymatic and non-enzymatic browning in cut fruits & vegetables and dried/dehydrated foods. Enzymatic browning is very common on the cut surfaces of fruits and vegetables, in which monophenols or diphenols are converted into brown colored polymeric quinones in presence of polyphenol oxidase. SO_2 reacts with the quinone intermediates and give substituted diphenols.

ii. Inhibit the growth of microorganisms (spoilage bacteria), yeast and molds by breaking the disulphide bonds of proteins and enzymes (of microorganisms), which are essential for their enzymatic activity (Fig. 15.9).

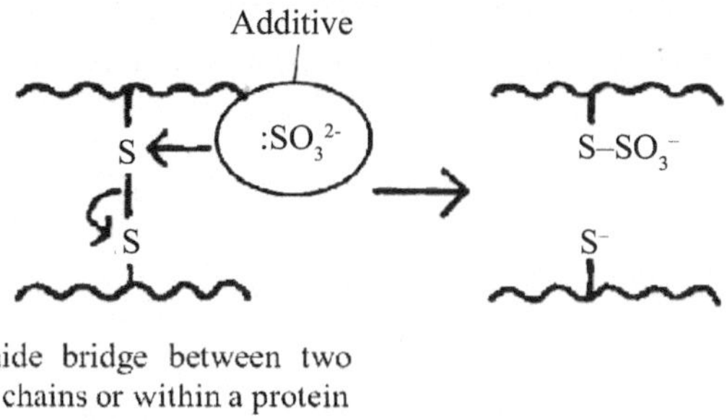

Fig. 15.9: Chemical action of SO_2 or other chemical forms on proteins present in microorganisms

iii. Reduce the oxidation of fats and other oxidative components in foods by deactivating ascorbic acid oxidase and lipoxygenase, for which the mechanism is not clear.

According to USFDA, SO_2 has ADI of 0-0.7mg/kg body weight, with maximum permissible level of 40-3000ppm in foods. At the levels used in foods, SO_2 has been listed as GRAS. However, it has been reported that some people might show the symptoms of nausea, diarrhea, breathing problems or even fatal shock after consuming SO_2added food products. Due to this reason, FDA has banned the addition of SO_2 when fruits and vegetables are presented in raw or fresh form to the consumers. Major applications of SO_2 are in the preservation of fruits and vegetables, dehydrated fruits, dried fruits, wine, etc.

10. Nitrites/nitrates: Sodium nitrates and sodium nitrites are chemical preservatives commonly used in the preservation of meat and fish products and during manufacturing of variety of cheeses. Nitrates and nitrites are naturally present in several foods such as vegetables, meat products, fruits, drinking water and cheese. Nitrates are also used as food additives for fixation of color of cured meat; and for the preservation of meat and fish products by inhibiting microbial growth particularly *Clostridium botulinum*, which has already been discussed in previous section 'curing agents'. When added in food products for preservation, nitrates are converted into nitrites and nitric oxide (NO). Nitric oxide further reacts with iron-sulphur proteins of bacteria and shows inhibitory action. It also inhibits the mitochondrial cytochrome oxidase and interrupts TCA cycle and other metabolic activities. For nitrates and nitrites, JECFA has prescribed the ADI 0-3.7 and 0-0.7mg/kg body weight/day, respectively.

Artificial sweeteners

Artificial sweeteners are sugar substitutes, which are used as an alternative of sugar to cut calories in diet or to manage diabetes and obesity. They provide zero to a few calories in diet and are many times sweeter than sugar. Till date,

FDA (USFDA, 2018) has approved six artificial sweeteners for use in food products at a specified level, which are mentioned below:

- Saccharin
- Aspartame
- Acesulfame potassium (Ace-K)
- Sucralose
- Neotame
- Advantame

Apart from the artificial sweeteners mentioned above, other sweeteners extracted from natural sources (with GRAS status) are given below.

- Steviol glycosides
- Luo Han Guo fruit extracts

1. Saccharine: 'Saccherine' derived from the Latin '*saccharum*' for sugar, was discovered in 1879. It is a **non-nutritive** sweetener approved by USFDA and is placed on the list of GRAS. Sacchrine is 200-700 times sweeter than table sugar (sucrose) depending upon its concentration. It does not provide any calories as it is not metabolized by the human body and excreted in the same form as ingested. It's parent compound is 1,2-benzoisothiazol-3-(2H), which is sparingly soluble in water. Therefore, its sodium or potassium salts are usually used as food additives.

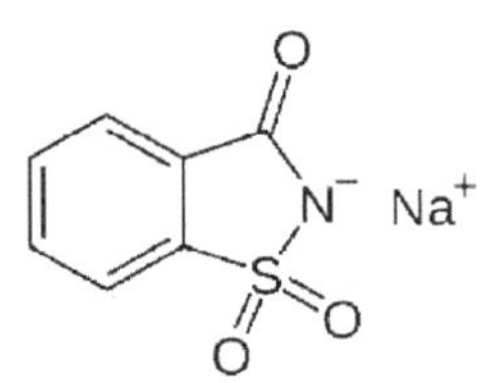

Saccharine

Sacchrine has slight metallic after taste and off-flavor. It is highly stable to heat and aqueous conditions over a wide range of pH. Therefore, it is suitable for several applications such as in fruit juices, beverages, processed fruits, chewing gum, sauces, desserts, etc. US FDA has prescribed its ADI 5mg/body weight/day. The ADI represents the amount of a food additive that can be consumed daily for a lifetime with no ill effect. It is not a maximum amount that can be safely consumed on a given day.

2. Aspartame: Aspartame, discovered in 1965, is a **nutritive** sweetener, approximately 200 times sweeter than sucrose. Chemically, aspartame is a dipeptide composed of two amino acids, L-aspartic acid and methyl ester of L-phenylalanine. As it is made up of amino acids, it provides calories equivalent to the proteins. In moist conditions, aspartame is not stable to heat and highly acidic conditions and loses its sweetness. In such conditions, ester bond is hydrolyzed and saccharine is degraded into aspartic acid, phenylalanine and methanol. It is stable in weak acidic range (pH 3-5) and highly stable in dried products. It is slightly soluble in water and not soluble in fats and oil. The taste of this sweetener is clean and sweet without after or metallic taste. Therefore, it is one of the most commonly used sweeteners in a variety of food products such as carbonated soft drinks, fruit juices, beverages, tabletop sweetener, powdered drinks, dairy products, dairy whiteners, etc. Due to its instability at higher temperatures, aspartame is not used in bakery products. US FDA has set its ADI 50 mg/kg body weight/day in humans and animals both.

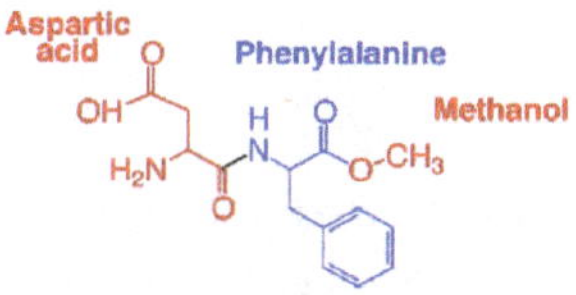

Aspartame

Owing to its safety in human population, US FDA has approved its use as a **'General purpose sweetener'.** However, patients who have lesser or no ability to digest phenylalanine (phynylketonurics), have difficulties with the consumption of aspartame. Another health concern with aspartame is the generation of methanol under physiological conditions.US FDA has stated that aspartame yields approximately 10% methanol by weight, which does not change the base line of methanol levels in blood. The amount of methanol released from aspartame is well below to the levels normally maintained in blood from the consumption of fruits, vegetables and juices.

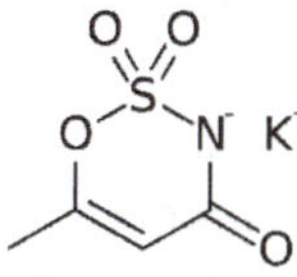

Acesulfame-K

3. Acesulfame potassium (Ace-K): Acesulfame potassium is a **non-nutritive** sweetener and approved by US FDA as a 'general purpose sweetener' except in meat and meat products. It is 130-200 times sweeter than sucrose. It is shelf stable over a long period and highly soluble in water. It is also stable to high temperatures over a wide range of pH. Therefore, it is safe to use in bakery products, beverages, fruit juices, desserts, confectionary products, flavored milk, ice-cream, etc.

It is also suitable to highly acidic low-calorie beverages (e.g. diet-cokes) due to its stability in aqueous solutions. Ace-K has not been reported to impart any off-taste after consumption. It is not metabolized by the human body and is excreted completely. US FDA has set its ADI 15mg/per kg body weight/day.

4. Sucralose: Sucralose is a **non-nutritive** sweetener and approved by US FDA as a **'General purpose sweetener'**. It is 600 times sweeter than sucrose and has pleasant taste with no after taste after consumption. The chlorination of sucrose molecule at 1 and 6 positions of fructose moiety and 4 position of glucose moiety produces highly sweetening and stable compound known as sucralose. It is highly soluble in water and heat stable over a wide range of pH. Sucralose is not hydrolyzed, dechlorinated or metabolized after ingestion; hence does not provide any calories. No toxic effects have been reported of sucralose consumption till date. It is safe to use in a variety of products such as frozen desserts, baked goods, dairy products, beverages, jam & jellies, confections, chewing gum, salad dressing, fillings, etc. Sucralose has ADI of 15 mg/kg body weight/day.

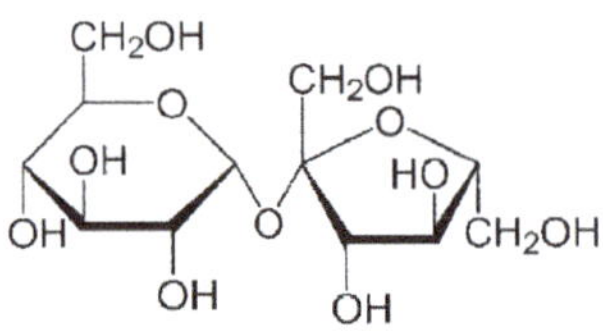

Sucrose

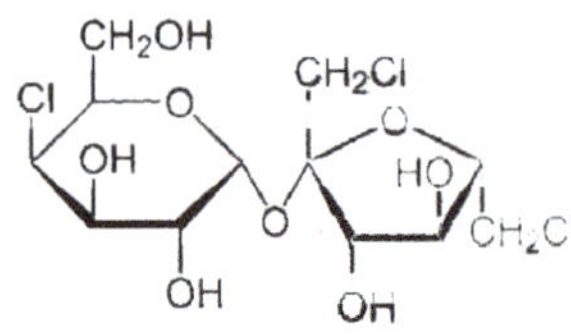

Sucralose

5. *Neotame:* Neotame is a new **non-nutritive** sweetener and flavor enhancer approved by US FDA as a 'General purpose sweetener', except in meat and meat products. It is approximately 7000-13000 times sweeter than sucrose depending upon the concentration. Structurally, it is similar to and derived from aspartame and is 30 to 60 times sweeter than aspartame. Chemically, neotame is N-[N-(3,3-dimethylbutyl)-l-α-aspartyl]-l-phenylalanine 1-methyl ester. It does not provide any calories after consumption. It has a clean sweet taste with no off-flavor or after taste. Unlike aspartame, it is stable in aqueous solutions, at high temperatures and over a wide range of pH and does not pose any problem to phenylketonurics. Neotame is most commonly used in carbonated soft drinks along with desserts, baked goods, beverages, fruit juices, dairy products, fillings, etc. Neotame has ADI of 18mg/kg body weight/day.

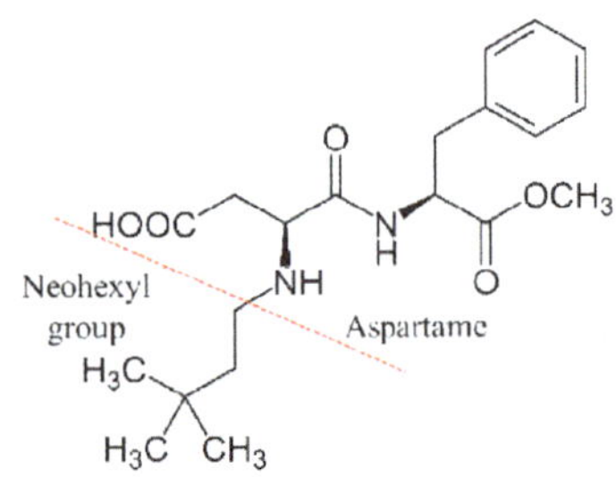

Neotame

6. *Advantame:* Advantame is a new non-nutritive sweetener, approved by USFDA in 2014 for use in food products except in meat and poultry. Chemically, it is N-[N-[3-(3-hydroxy-4-methoxy-phenyl) propyl]-[α]-aspartyl]-L-phenylalanine 1-methyl ester, which is approximately 20,000 times sweeter than sucrose. It imparts clean sweet taste like aspartame. It is stable to high temperatures and a range of pH; and can be used in a number of food products such as non-alcoholic beverages, carbonated soft drinks along with desserts, baked goods, beverages, fruit juices, dairy products, fillings, etc. JECFA has set its ADI 32.8 mg/kg body weight/day.

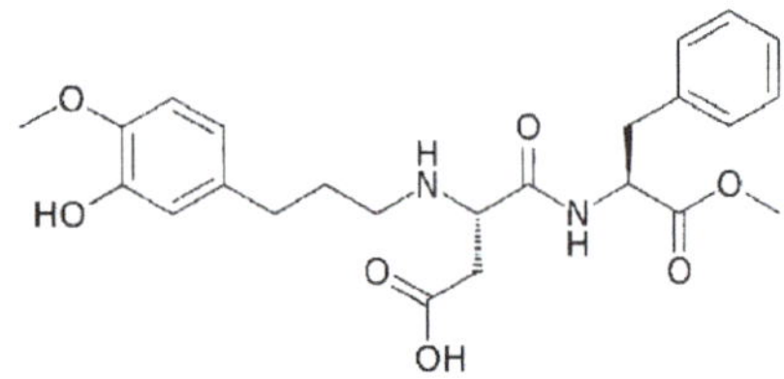

Advantame

Natural alternative sweeteners

1. Steviol glycosides: Steviol glycosides are natural constituents of the leaves of *Stevia rebaudiana*, which is a plant native to South America and commonly known as Stevia. They are **non-nutritive** sweeteners with GRAS status, but not FDA approved yet. It should be noted that the additives which have GRAS status, do not require the approval of USFDA. Steviol glycosides are approximately 200 to 400 times sweeter than table sugar. It is not metabolized by human body; therefore, it does not provide any calories.

Stevia plant

A few reports have shown slight bitter and after taste after consumption of stevia added products. In market, stevia leaf extract or leaves powder are also available as alternative sweeteners, but these forms are not placed on the list of GRAS additives. JECFA has established the ADI of pure steviol glycosides to 4 mg/kg body weight/day.

2. Luo Han Guo fruit extracts: Siraitia grosvenorii Swingle fruit extract (SGFE) contains varying levels of mogrosides, which are the non-nutritive constituents of the fruit primarily responsible for the characteristic sweetness of SGFE. SGFE, depending on the mogroside content, is reported to be 100 to 250 times sweeter than sugar. *Siraitia grosvenorii* Swingle, commonly known as Luo Han Guo or monk fruit, is a plant native to Southern China. Although it has gained the GRAS status, JECFA has not specified it ADI for human consumption yet.

Luo Han Guo fruit plant and dried fruits

Natural food toxins

Natural toxins are toxic compounds produced from the living organisms such as plants, microorganisms or animals. These toxins are not harmful to the organisms from which they are produced, but may be toxic to the other organisms. Several toxic components are found in plants such as phytates, solanin, cyanogenic glycosides, alkaloids, etc. which are also known as ant nutritional factors. These factors play a defensive role for the protection of plants by pests, birds, animals or even humans. In foods, toxic elements can be found naturally as a part of defence mechanism or through their infestation with toxin-producing molds. These toxins when consumed through foods can exert mild to adverse health effects. They can be neurotoxic, hypersensitive, allergenic or cytotoxic in nature and can show the symptoms of headache, severe stomachache, irritability, diarrhea and even death. Several food toxins, their source and action are summaries below in Table 15.6.

Adverse health effects of natural toxins

- General toxicity (ill effects and even death)
- Mutagenic effects
- Carcinogenic effects
- Teratogenic effects (Impairs with the development of fetus or embryo)
- Hormonal imbalance

Table 15.6: Food toxins, their source and action

Class	Toxin	Source	Functions
Cyanogenic glycosides	Amygdalin	Bitter apricot kernels, bitter almonds, peach kernel, apple, cheery, peach, plum	Cause cyanide poisoning, interrupts with the activity of cytochrome oxidase in electron transport chain and cause no ATP formation
	Dhurrin	Sorghum, Kaffir corns	
	Linamarin	Lima beans, flaxseeds, cassava	
	Prunasin	Bitter apricot kernels, bitter almonds, peach kernel, apple, cheery, peach, plum	
	Prussic acid (HCN)	Cheery, apple, apricot, peach	
Lectins	Phytohaemagglutinin	Black beans, soybeans, lima beans, kidney beans and lentils, grains	Bind with mucosal cells and interrupts with the nutrients absorption, induce mitosis, affect membrane transport and permeability to proteins, agglutinate red blood cells
Anti-vitamin compounds	Thiaminase	Fish, crab, clams and in some fruits and vegetables such as blueberries, black currants, red beets, brussels sprouts and red cabbage	Thiaminase cleaves thiamin and impairs the carbohydrate metabolism and energy production, may cause to 'beri-beri' disease
Alkaloids	Pyrrolizidine alkaloids	Members of Apocyanacae, legumes, faba beans, etc.	Hepatotoxic, mutagenic, teratogenic and carcinogenic
	Solanine	Potato, eggplant, apples, bell peppers, cherries, sugar beets and tomatoes	Inhibitors of acetylcholinesterase and disruptors of cell membranes. Symptoms include drowsiness, itchiness in the neck region, increased sensitivity, labored breathing, abdominal pain, nausea, vomiting and diarrhea
Enzymes inhibitor	Amylase inhibitor	Wheat, rye and kidney beans	Inhibits the alpha-amylase activity and interfere with starch digestion, absorption and assimilation

(Contd.)

Furocoumarins	Psoralen,5-methoxypsoralen and 8-methoxypsoralen	Citrus fruits, parsnip, parsley, celery, carrots	It forms addition products (adducts) with DNa and DNA cross links and cause mutations, chromosome aberrations and cell death. Also shows the toxicity to the skin
-	Erucic acid	Rapeseed	Growth retardation, thickening of the epicardium, interruption with contraction of cardiac muscles
Goitrogens	Glucosinolates	Spinach, cassava, peanuts, soybeans, broccoli, brussels sprouts, cabbage, canola, cauliflower, mustard, rapeseed	Adversely affect the growth, cognitive functions and hormonal balance by interfering with the uptake of iodine, cause goiter.

Radionuclides in foods

Radionuclides are the radioactive isotopes which continuously emits radiations. These are naturally present in all foods, water and air and the environment. In foods, they can come from atmosphere, soil or water depending on the geology, agro-climatic conditions and agricultural practices. We are exposed to radiation (also known as background radiation) from these radionuclides on a daily basis. The common radio nuclides present in food are potassium-40 (^{40}K), radium-226 (^{226}Ra) and uranium-238 (^{238}U). ^{40}K is the most common naturally occurring radionuclide present in foods and dairy products. For example, milk contains approximately 50 Bq/L ^{40}K. Other radio nuclides such as uranium and thorium are also present in foods and milk but in very lower amount. When large amount of radioisotopes are released in environment either from nuclear industry or from nuclear fission, they enter in to the water, soil and air; from which they are taken up by the plants and animals along with the nutrients.

Commodities of concern

Open air fruits and vegetables, plants, wild animal meat, mushroom, berries, etc. can be the major concerns for radio nuclides contamination. Fish and other aquatic animals also accumulate the radio nuclides. Milk is also associated with radionuclide contamination due to transfer of such elements from soil to feed to animal. Radioactive iodine (I^{131}) is of major concern in milk and milk products due to rapid transfer of it from feed to animal, which further accumulates in thyroid gland and increase the incidence of thyroid cancer.

Possible health effects

- **Acute health effects:** Nausea, skin irritation, vomiting, fatigue, headache, extreme tiredness, hair loss, and organs failure
- **Chronic health effects:** Increased risk of thyroid cancer, skin cancer

Sample questions

1. Define food additives.
2. Why food additives are added in food products? Discuss in detail.
3. Define antioxidants with examples.
4. Explain the mechanism of action of antioxidant in prevention of oxidation of fat.
5. Enlist natural and synthetic antioxidants.
6. Describe the action of chelating agents. How chelating agents are different from antioxidants?

7. What are the roles of chelating agents in foods?
8. Give detailed classification of colorants along with the purposes of addition of colorants in foods.
9. Explain the role of curing agents in foods.
10. Discuss the role of sodium nitrite in curing of meat and meat products.
11. Differentiate among Natural flavorings, Artificial flavorings and Nature-identical flavorings.
12. Write the short notes on:
    a. Flavor enhancers
    b. Purine 5'-Ribonucleotides as flavor enhancer
    c. Sulphur dioxide as food preservative
    d. Advantame
    e. Luo Han Guo fruit extracts
13. Discuss various food preservatives, their mode of action and food applications in detail.
14. Write a short note on natural food toxins.

Suggested Readings

Aamoth, H.L. and Butt, F.J. (1960). Maintaining food quality with chelating agents. *Annals of the New York Academy of Sciences*, 88(1): 526-531.

Ashurst, P.R., (2012). *Food flavorings*. Springer Science & Business Media.

Chattopadhyay, S., Raychaudhuri, U. and Chakraborty, R. (2014). Artificial sweeteners–a review. *Journal of Food Science and Technology*, 51(4):611-621.

EFSA Panel on Food additives and Nutrient Sources added to Food (ANS), (2014). Scientific Opinion on the re evaluation of propionic acid (E 280), sodium propionate (E 281), calcium propionate (E 282) and potassium propionate (E 283) as food additives. *EFSA Journal*, 12(7):3779.

Gray, J.I. and Pearson, A.M. (1984). Cured meat flavor. In *Advances in food research* (Vol. 29, pp. 1-86). Academic Press.

Manay, N.S.O. (2001). Food: facts and principles. New Age International.

Martins, N., Roriz, C.L., Morales, P., Barros, L. and Ferreira, I.C. (2016). Food colorants: Challenges, opportunities and current desires of agro-industries to ensure consumer expectations and regulatory practices. *Trends in Food Science & Technology*, 52:1-15.

US FDA (2018). Additional Information about High-Intensity Sweeteners Permitted for Use in Food in the United States.

Wedzicha, B.L. (1984) Sulphur dioxide in foods:–chemical interactions. *Nutrition Bulletin*, 9(3): 155-164.

https://www.fda.gov/Food/IngredientsPackagingLabeling/FoodAdditivesIngredients/ucm397725.htm

FSC
www.fsc.org
MIX
Papier | Fördert gute Waldnutzung
FSC® C083411

Zeitfracht Medien GmbH
Ferdinand-Jühlke-Straße 7
99095 Erfurt, Deutschland
produktsicherheit@kolibri360.de